Busy Ant Maths

Stretch and Challenge 5

A problem-solving, cross-curricular programme for children working above end-of-year expectations

Peter Clarke

William Collins' dream of knowledge for all began with the publication of his first book in 1819. A self-educated mill worker, he not only enriched millions of lives, but also founded a flourishing publishing house. Today, staying true to this spirit, Collins books are packed with inspiration, innovation and practical expertise. They place you at the centre of a world of possibility and give you exactly what you need to explore it.

Collins. Freedom to teach.

Published by Collins

An imprint of HarperCollinsPublishers
The News Building
1 London Bridge Street
London SE1 9GF

HarperCollins*Publishers*
1st Floor, Watermarque Building, Ringsend Road
Dublin 4, Ireland

Browse the complete Collins catalogue at
www.collins.co.uk

10 9 8 7 6 5 4

ISBN 978-0-00-816734-9

British Library Cataloguing in Publication Data
A Catalogue record for this publication is available from the British Library.

Commissioned by Fiona McGlade
Cover and series design by Kneath Associates
Template creation by Ken Vail Graphic Design
Illustrations by Mark Walker, Steve Evans, Gwyneth Williamson and QBS Learning
Typesetting by QBS Learning
Editing and proofreading by Alissa McWhinnie
Printed and bound in the UK using 100% Renewable
Electricity at CPI Group (UK) Ltd

Acknowledgements
Peter Clarke wishes to thank Brian Molyneaux for his valuable contribution to this publication.

Contents

Quick reference guide to *Stretch and Challenge 5*

Domain(s)	Topic	Issue number	Teacher's notes page number
Number: – Number and place value	Number	1	98
	Number	2	101
	Number	3	104
	Negative numbers	4	107
Number: – Addition and subtraction	Addition	5	110
	Subtraction	6	113
Number: – Multiplication and division	Multiplication	7	116
	Division	8	119
Number: – Addition and subtraction – Multiplication and division	Mixed operations	9	122
	Mixed operations	10	125
	Mixed operations	11	128
	Mixed operations	12	131
Number: – Fractions (including decimals and percentages)	Fractions	13	135
	Fractions	14	138
	Decimals	15	141
	Percentages	16	144
	Percentages	17	147
	Fractions, decimals and percentages	18	150
Measurement	Length	19	153
	Mass	20	156
	Volume and capacity	21	160
	Time	22	163
	Temperature	23	166
	Area	24	169
	Area and perimeter	25	172
	Measurement	26	175
Geometry: – Properties of shapes	2-D shapes	27	178
	3-D shapes	28	181
	Movement and angle	29	184
Geometry: – Position and direction	Reflections	30	187
	Translations	31	190
Geometry: – Properties of shapes – Position and direction	Reflective and rotational symmetry	32	193
	Geometry	33	196
	Geometry	34	199
Statistics	Statistics	35	202
	Statistics	36	205

Introduction

The National Curriculum emphasises the importance of all children mastering the programme of study taught each year and discourages the acceleration of children into content from subsequent years.

The National Curriculum states: *'The expectation is that the majority of pupils will move through the programmes of study at broadly the same pace. However, decisions about when to progress should always be based on the security of pupils' understanding and their readiness to progress to the next stage. Pupils who grasp concepts rapidly should be challenged through being offered rich and sophisticated problems before any acceleration through new content. Those who are not sufficiently fluent with earlier material should consolidate their understanding, including through additional practice, before moving on.'* [1]

However, the National Curriculum also goes on to say that: *'Within each key stage, schools [therefore] have the flexibility to introduce content earlier or later than set out in the programme of study. In addition, schools can introduce key stage content during an earlier key stage, if appropriate.'* [2]

Stretch and Challenge aims to provide support in meeting the needs of those children who are exceeding age-related expectations by providing a range of problem-solving and cross-curricular activities designed to enrich and deepen children's mathematical knowledge, skills and understanding.

The series provides opportunities for children to reason mathematically and to solve increasingly complex problems, doing so with fluency, as described in the aims of the National Curriculum:

'The National Curriculum for mathematics aims to ensure that all pupils:

- *become **fluent** in the fundamentals of mathematics, including through varied and frequent practice with increasingly complex problems over time, so that pupils develop conceptual understanding and the ability to recall and apply knowledge rapidly and accurately*
- ***reason mathematically** by following a line of enquiry, conjecturing relationships and generalisations, and developing an argument, justification or proof using mathematical language*
- *can **solve problems** by applying their mathematics to a variety of routine and non-routine problems with increasing sophistication, including breaking down problems into a series of simpler steps and persevering in seeking solutions.'* [3]

Stretch and Challenge has been designed to provide:

- a flexible 'dip-in' resource that can easily be adapted to meet the needs of individual children, and different classroom and school organisational arrangements
- enrichment activities that require children to use and apply their mathematical knowledge, skills and understanding to reason mathematically and to solve increasingly complex problems
- mathematical activities linked to the entire primary curriculum, thereby ensuring a range of cross-curricular contexts
- an easy-to-use bank of activities to save teachers time in thinking up new enrichment activities
- an interesting, unique and consistent approach to presenting enrichment activities to children.

The *Stretch and Challenge* series consists of six packs, also available digitally on Collins Connect, one for each year group from Year 1 to Year 6.

---

1 Mathematics programmes of study: key stages 1 and 2 National Curriculum in England, September 2013, page 3

2 Mathematics programmes of study: key stages 1 and 2 National Curriculum in England, September 2013, page 4

3 Mathematics programmes of study: key stages 1 and 2 National Curriculum in England, September 2013, page 3

Printed resources

Containing:

- Pupil activity booklets (Issues)
- Teacher's notes
- Resource sheets

Online resources at connect.collins.co.uk

Containing printable:

- Pupil activity booklets (Issues)
- Teacher's notes
- Resource sheets

It is envisaged that the activities in *Stretch and Challenge* will be used by either individuals or pairs of children. However, given the flexible nature of the resource, if appropriate, children can work in groups. The activities are intended to be used:

- as additional work to be done once children have finished other set work
- by those children who grasp concepts rapidly and need to be challenged through rich and sophisticated problems
- as in-depth work that is to be undertaken over a prolonged period of time, such as during the course of several lessons, a week or a particular unit of work
- as a resource for promoting mathematical reasoning and problem solving and developing independent thinking and learning
- as a springboard for further investigations into mathematics based on the children's suggestions.

The features of *Stretch and Challenge*

Pupil activity booklets (Issues)

- Each of the 36 Issues in *Stretch and Challenge 5* consists of a four-page A5 pupil activity booklet (to be printed double sided onto one sheet of A4 paper).
- The 36 Issues cover the different domains and attainment targets of the Mathematics National Curriculum Programme of Study (see pages 11–20).
- The Issues have been designed to resemble a newspaper, with each of the Issues consisting of between five and eight different activities, all related to the same mathematical topic.
- It is important to note that children are not expected to complete all the activities in an Issue nor work their way through an Issue from beginning to end. For many children not all of the activities offered in an Issue will be appropriate. When choosing which activities a child is to complete, teachers should ensure that the activities do not accelerate the child into mathematical content they may not be familiar with, or are unable to reason more deeply in order to develop a conceptual understanding. Rather, activities should be chosen on the basis that they engage the child in reasoning and the development of mathematical thinking, as well as enriching and deepening the child's mathematical knowledge, skills and understanding.
- The terms 'Issue' and 'Volume' have been used rather than 'Unit' and 'Year group' because they are in keeping with the newspaper theme.

Types of activities

- Each of the 36 Issues in *Stretch and Challenge 5* are designed to deepen children's mathematical knowledge, skills and understandings, and enhance their use and application of mathematics. There are four different types of 'using and applying' activities in the series:

  What's the Problem? | The Puzzler

  Looking for Patterns | Let's Investigate

- Alongside developing children's problem-solving skills, the series also provides activities with cross-curricular links to other subjects in the primary curriculum. The following shows the *Stretch and Challenge* features and its corresponding primary curriculum subject.

Curriculum subject	*Stretch and Challenge* feature
English	The Language of Maths
Science	Focus on Science
Computing	Technology Today
Geography	Around the World
History	In the Past
	Famous Mathematicians
Art and design / Music	The Arts Roundup
Design and Technology	Construct
Physical Education	Sports Update

- As well as the features mentioned above, other regular features in *Stretch and Challenge* include:

  Money Matters | At Home (home–school link activities).

- A chart showing the link between the Issues, the *Stretch and Challenge* features and cross-curricular links can be found on pages 21 and 22.
- Inquisitive ant is a recurring feature of the series. In each Issue there is an ant holding a mathematical word or symbol. Children locate the ant and write about the meaning of the word or symbol.

Teacher's notes

Each of the 36 Issues includes a set of teacher's notes, including answers.

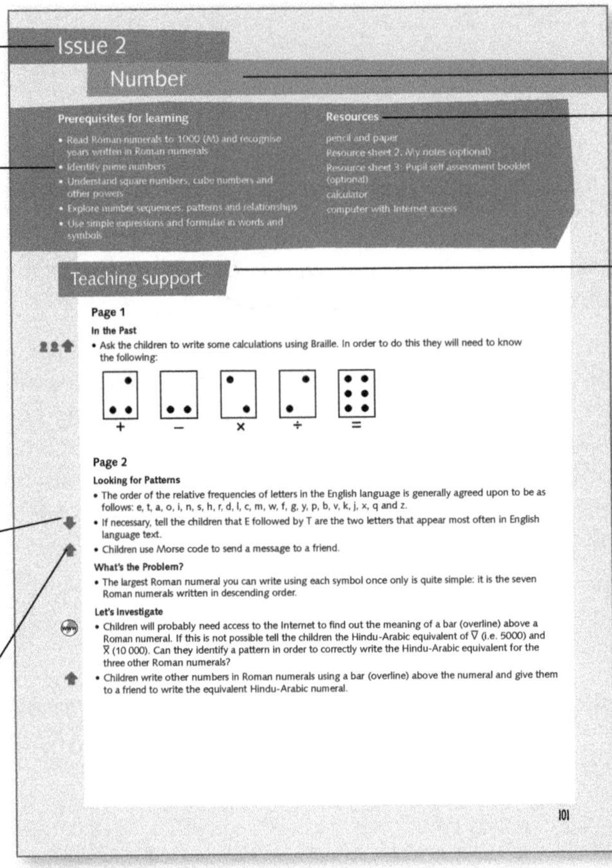
Issue 2

Number

Prerequisites for learning

- Read Roman numerals to 1000 (M) and recognise years written in Roman numerals
- Identify prime numbers
- Understand square numbers, cube numbers and other powers
- Explore number sequences, patterns and relationships
- Use simple expressions and formulae in words and symbols

Resources

pencil and paper
Resource sheet 2: My notes (optional)
Resource sheet 3: Pupil self assessment booklet (optional)
calculator
computer with Internet access

Teaching support

Page 1

In the Past

- Ask the children to write some calculations using Braille. In order to do this they will need to know the following:

\+ − × ÷ =

Page 2

Looking for Patterns

- The order of the relative frequencies of letters in the English language is generally agreed upon to be as follows: e, t, a, o, i, n, s, h, r, d, l, c, m, w, f, g, y, p, b, v, k, j, x, q and z.
- If necessary, tell the children that E followed by T are the two letters that appear most often in English language text.
- Children use Morse code to send a message to a friend.

What's the Problem?

- The largest Roman numeral you can write using each symbol once only is quite simple: it is the seven Roman numerals written in descending order.

Let's Investigate

- Children will probably need access to the Internet to find out the meaning of a bar (overline) above a Roman numeral. If this is not possible tell the children the Hindu-Arabic equivalent of $\overline{V}$ (i.e. 5000) and $\overline{X}$ (10 000). Can they identify a pattern in order to correctly write the Hindu-Arabic equivalent for the three other Roman numerals?
- Children write other numbers in Roman numerals using a bar (overline) above the numeral and give them to a friend to write the equivalent Hindu-Arabic numeral.

101

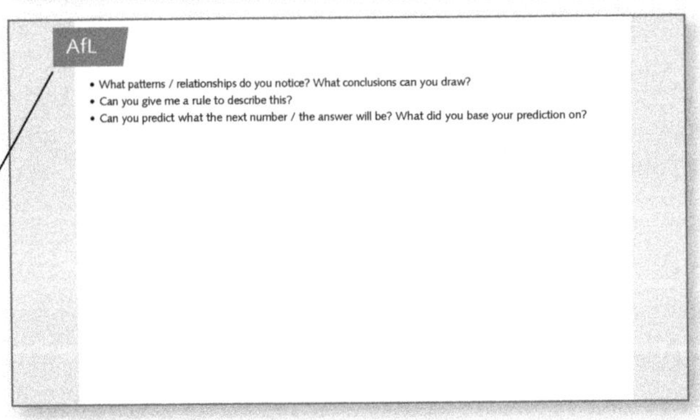
AfL

- What patterns / relationships do you notice? What conclusions can you draw?
- Can you give me a rule to describe this?
- Can you predict what the next number / the answer will be? What did you base your prediction on?

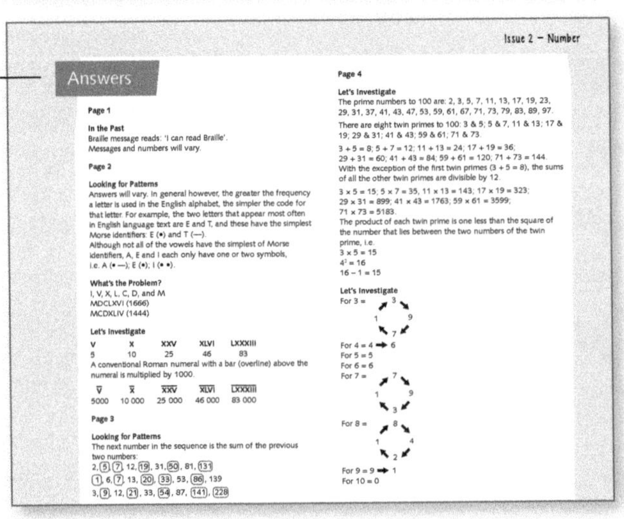
Issue 2 – Number

Answers

Page 1

In the Past
Braille message reads: 'I can read Braille'.
Messages and numbers will vary.

Page 2

Looking for Patterns
Answers will vary. In general however, the greater the frequency a letter is used in the English alphabet, the simpler the code for that letter. For example, the two letters that appear most often in English language text are E and T, and these have the simplest Morse identifiers: E (•) and T (—).
Although not all of the vowels have the simplest of Morse identifiers, A, E and I each only have one or two symbols, i.e. A (• —); E (•); I (• •).

What's the Problem?
I, V, X, L, C, D, and M
MDCLXVI (1666)
MCDXLIV (1444)

Let's Investigate

V	X	XXV	XLVI	LXXXIII
5	10	25	46	83

A conventional Roman numeral with a bar (overline) above the numeral is multiplied by 1000.

$\overline{V}$	$\overline{X}$	$\overline{XXV}$	$\overline{XLVI}$	$\overline{LXXXIII}$
5000	10 000	25 000	46 000	83 000

Page 3

Looking for Patterns
The next number in the sequence is the sum of the previous two numbers:
2, [5], [7], 12, [19], 31, [50], 81, [131]
[1], 6, [7], 13, [20], [33], 53, [86], 139
3, [9], 12, [21], 33, [54], 87, [141], [228]

Page 4

Let's Investigate
The prime numbers to 100 are: 2, 3, 5, 7, 11, 13, 17, 19, 23, 29, 31, 37, 41, 43, 47, 53, 59, 61, 67, 71, 73, 79, 83, 89, 97.
There are eight twin primes to 100: 3 & 5; 5 & 7, 11 & 13; 17 & 19; 29 & 31; 41 & 43; 59 & 61; 71 & 73.
3 + 5 = 8; 5 + 7 = 12; 11 + 13 = 24; 17 + 19 = 36; 29 + 31 = 60; 41 + 43 = 84; 59 + 61 = 120; 71 + 73 = 144.
With the exception of the first twin primes (3 + 5 = 8), the sums of all the other twin primes are divisible by 12.
3 × 5 = 15; 5 × 7 = 35, 11 × 13 = 143; 17 × 19 = 323; 29 × 31 = 899; 41 × 43 = 1763; 59 × 61 = 3599; 71 × 73 = 5183.
The product of each twin prime is one less than the square of the number that lies between the two numbers of the twin prime, i.e.
3 × 5 = 15
$4^2 = 16$
16 − 1 = 15

Let's Investigate
For 3 = 3 → 9 → 7 → 1
For 4 = 4 → 6
For 5 = 5
For 6 = 6
For 7 = 7 → 9 → 3 → 1
For 8 = 8 → 4 → 2 → 1
For 9 = 9 → 1
For 10 = 0

Issue number

Prerequisites for learning

Lists the prerequisites for learning that children need to have acquired prior to this Issue.

Lists the associated knowledge and skills that contribute to understanding the Issue topic.

Simplifications

Where appropriate, offers suggestions for supporting children who may be experiencing difficulties understanding the main mathematical ideas.

Extensions

Where appropriate, offers suggestions for extending children's understanding if you feel they are developing a good understanding of the main mathematical ideas.

Assessment for Learning

Each Issue includes a list of questions specifically designed to assist in assessing pupils' understanding of the Issue topic.

Answers

These are provided where appropriate.

Mathematics topic

Resources

To aid preparation, the resources needed for the Issue are listed.

Teaching support

Provides teaching points for each of the activities in the Issue. These may be helpful when introducing the Issue to the children, or when children experience difficulty whilst working on a particular activity.

Almost all of the activities in *Stretch and Challenge* can be undertaken either individually or in pairs (or sometimes in small groups).

Where an activity is particularly suitable for pairs to work on, this is denoted by 👥.

When children are likely to need access to a computer, this is denoted by 💻.

When children are likely to need access to a computer with Internet access, this is denoted by (www).

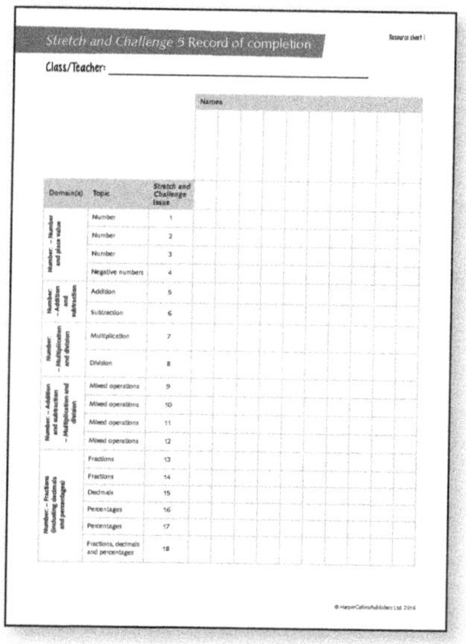

Stretch and Challenge 5 Record of completion

Resource sheet 1

Class/Teacher: ____________________

Domain(s)	Topic	Stretch and Challenge Issue	Names
Number – Number and place value	Number	1	
	Number	2	
	Number	3	
	Negative numbers	4	
Number – Addition and subtraction	Addition	5	
	Subtraction	6	
Number – Multiplication and division	Multiplication	7	
	Division	8	
Number – Addition and subtraction – Multiplication and division	Mixed operations	9	
	Mixed operations	10	
	Mixed operations	11	
	Mixed operations	12	
Number – Fractions (including decimals and percentages)	Fractions	13	
	Fractions	14	
	Decimals	15	
	Percentages	16	
	Percentages	17	
	Fractions, decimals and percentages	18	

© HarperCollinsPublishers Ltd 2016

Record of completion

To assist in keeping a record of which Issues children have completed.

Once a child has completed an Issue you could either put a tick or write the date in the corresponding box.

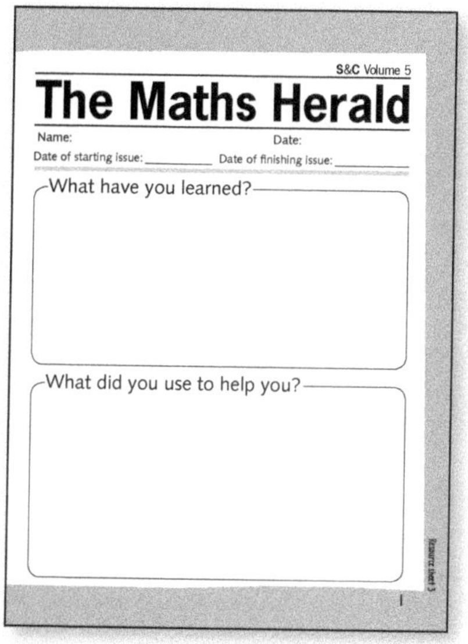

S&C Volume 5

The Maths Herald

Name: Date:

Date of starting issue: ________ Date of finishing issue: ________

What have you learned?

What did you use to help you?

Resource sheet 3

Pupil self assessment booklet

Each Resource Pack in the *Stretch and Challenge* series includes an age-appropriate pupil self assessment A5 booklet (to be printed double sided onto one sheet of A4 paper).

This booklet is a generic sheet that can be used for any, or all, of the 36 Issues in the Resource Pack.

The booklet is designed to provide children with an opportunity to undertake some form of self assessment once they have completed the Issue.

After the children have completed the booklet, discuss with them what they have written.

This can then be kept, together with the child's copy of the Issue and their working out and answers, including, if appropriate, 'My notes'.

S&C Volume 5

The Maths Herald

Name: Date:

Date of starting issue: ________ Date of finishing issue: ________

My notes

Resource sheet 2

My notes

The pupil activity booklets have been designed to resemble a newspaper. This means that quite often there is insufficient space in the booklets for children to show their working and answers.

You may decide to simply provide children with pencil and paper to record their work or an exercise book that they use as their '*Stretch and Challenge* Journal'. Alternatively, you could provide them with a copy of the A5 booklet: 'My notes' (to be printed doubled sided onto one sheet of A4 paper). This can then be kept, together with the child's copy of the Issue and, if appropriate, their 'Pupil self assessment booklet.'

Whichever method you choose for the children to record their working and answers, i.e. on sheets of paper, using a 'My notes' booklet, in an exercise book, or any other method, children need to be clear and systematic in their recording.

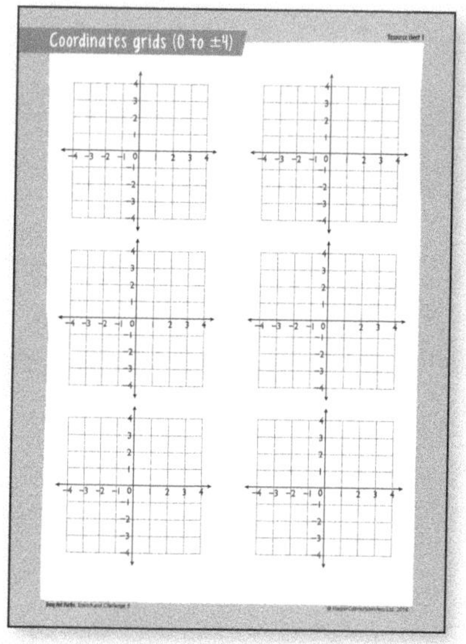

Other Resource sheets

For some of the activities, children are required to use a specific Resource sheet.

These are included both in the back of this Resource Pack and the online resources.

A possible *Stretch and Challenge* teaching and learning sequence

As the diagram on the right illustrates, the process of learning about mathematics can be thought of as the interrelationship between knowledge, understanding and application.

A suggested teaching sequence for working with children based on this model and using the activities in *Stretch and Challenge* is given below.

A complete sequence may occur during a particular lesson if the activity given is designed to be completed during the course of the lesson. Alternatively, the teaching and learning sequence may extend for a longer period of time if the activities are to be completed over the course of several lessons, a week or during a particular unit of work.

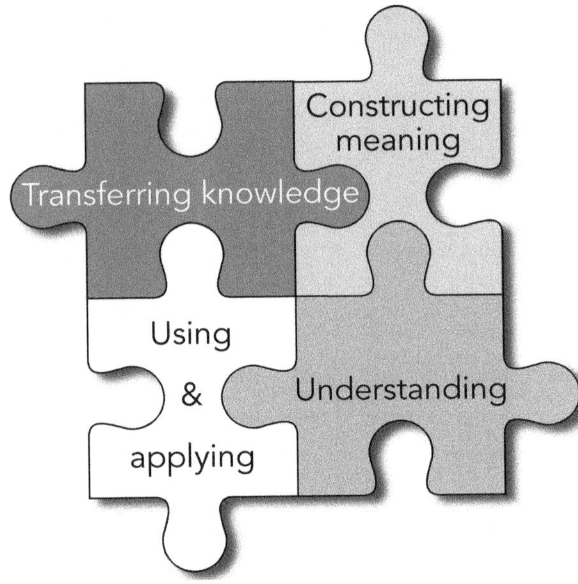

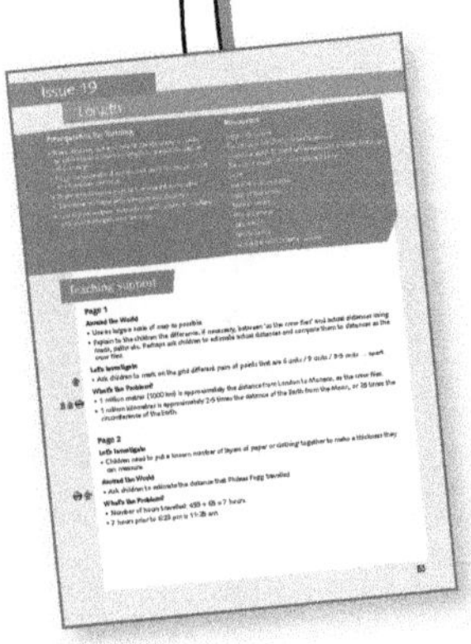

Briefing

The teacher:

- introduces the topic
- checks prerequisites for learning
- introduces the activity including, where appropriate, reading through the activity with the children
- checks for understanding
- clarifies any misconceptions
- ensures there is easy access to any necessary resources.

Working

- Children work individually, in pairs, or if appropriate, in groups.
- Teacher acts only as a 'guide-on-the-side'.

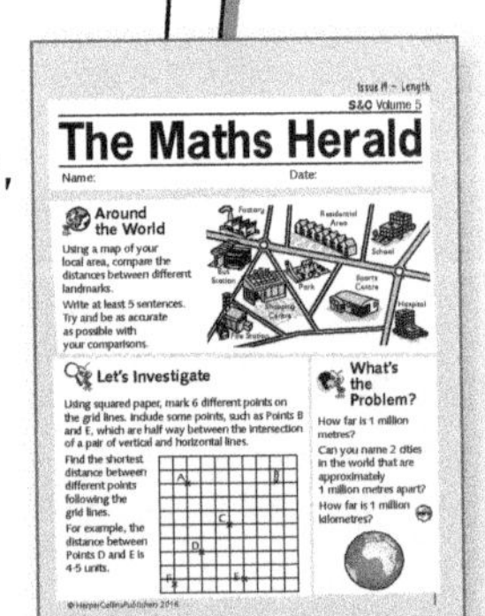

De-briefing

The child / children:

- reports back to others
- reflects on their learning
- identifies the 'next step'.

The teacher evaluates learning.

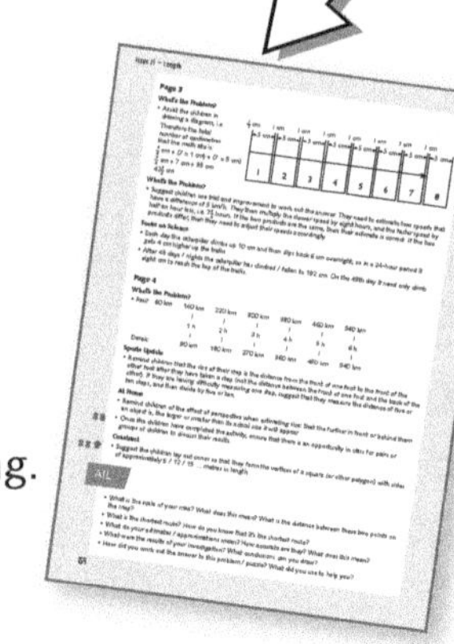

Links to the Year 5 Mathematics National Curriculum Programme of Study and Attainment Targets

	Stretch and Challenge Issue 1	2	3	4	5	6	7	8	9	10	11	12	13	14	15	16	17	18	19	20	21	22	23	24	25	26	27	28	29	30	31	32	33	34	35	36
Number – Number and place value																																				
• read, write, order and compare numbers to at least 1 000 000 and determine the value of each digit	•	•	•		•	•	•	•	•	•	•	•																								
• count forwards or backwards in steps of powers of 10 for any given number up to 1 000 000		•	•																																	
• interpret negative numbers in context, count forwards and backwards with positive and negative whole numbers, including through zero				•																																
• round any number up to 1 000 000 to the nearest 10, 100, 1000, 10 000 and 100 000																																				
• solve number problems and practical problems that involve all of the above	•	•	•	•	•	•	•	•	•	•	•	•																								
• read Roman numerals to 1000 (M) and recognise years written in Roman numerals		•	•									•																								
Notes and guidance (non-statutory)																																				
Pupils identify the place value in large whole numbers.																																				
They continue to use number in context, including measurement. Pupils extend and apply their understanding of the number system to the decimal numbers and fractions that they have met so far.	•	•	•	•	•	•	•	•	•	•	•	•																								
They should recognise and describe linear number sequences, including those involving fractions and decimals, and find the term-to-term rule.	•	•	•																																	
They should recognise and describe linear number sequences (for example, 3, $3\frac{1}{2}$, 4, $4\frac{1}{2}$ ...), including those involving fractions and decimals, and find the term-to-term rule in words (for example, add $\frac{1}{2}$).																																				

	Stretch and Challenge Issue																																			
	1	2	3	4	5	6	7	8	9	10	11	12	13	14	15	16	17	18	19	20	21	22	23	24	25	26	27	28	29	30	31	32	33	34	35	36
Number – Addition and subtraction																																				
• add and subtract whole numbers with more than 4 digits, including using formal written methods (columnar addition and subtraction)					•	•				•	•	•																								
• add and subtract numbers mentally with increasingly large numbers				•	•	•			•	•	•	•																								
• use rounding to check answers to calculations and determine, in the context of a problem, levels of accuracy										•		•																								
• solve addition and subtraction multi-step problems in contexts, deciding which operations and methods to use and why					•	•			•	•	•	•																								
Notes and guidance (non-statutory)																																				
Pupils practise using the formal written methods of columnar addition and subtraction with increasingly large numbers to aid fluency					•	•				•	•	•																								
They practise mental calculations with increasingly large numbers to aid fluency (for example, 12 462 – 2300 = 10 162).				•	•	•			•	•	•	•																								

Stretch and Challenge Issue

Number – Multiplication and division	1	2	3	4	5	6	7	8	9	10	11	12	13	14	15	16	17	18	19	20	21	22	23	24	25	26	27	28	29	30	31	32	33	34	35	36
• identify multiples and factors, including finding all factor pairs of a number, and common factors of two numbers	•	•	•				•					•																								
• know and use the vocabulary of prime numbers, prime factors and composite (non-prime) numbers		•	•										•																							
• establish whether a number up to 100 is prime and recall prime numbers up to 19		•	•										•																							
• multiply numbers up to 4 digits by a one- or two-digit number using a formal written method, including long multiplication for two-digit numbers							•			•		•																								
• multiply and divide numbers mentally drawing upon known facts							•	•	•	•	•	•																								
• divide numbers up to 4 digits by a one-digit number using the formal written method of short division and interpret remainders appropriately for the context								•		•		•																								
• multiply and divide whole numbers and those involving decimals by 10, 100 and 1000																																				
• recognise and use square numbers and cube numbers, and the notation for squared (^{2}) and cubed (^{3})	•	•	•			•	•			•	•																•									
• solve problems involving multiplication and division including using their knowledge of factors and multiples, squares and cubes	•	•	•				•	•	•	•	•	•																								
• solve problems involving addition, subtraction, multiplication and division and a combination of these, including understanding the meaning of the equals sign						•	•	•	•	•	•	•																								
• solve problems involving multiplication and division, including scaling by simple fractions and problems involving simple rates							•	•	•	•	•	•																								

	Stretch and Challenge Issue																																			
Notes and guidance (non-statutory)	1	2	3	4	5	6	7	8	9	10	11	12	13	14	15	16	17	18	19	20	21	22	23	24	25	26	27	28	29	30	31	32	33	34	35	36
Pupils practise and extend their use of the formal written methods of short multiplication and short division. They apply all the multiplication tables and related division facts frequently, commit them to memory and use them confidently to make larger calculations.							•	•		•	•	•																								
They use and understand the terms factor, multiple and prime, square and cube numbers.	•	•	•			•	•	•		•	•	•	•														•									
Pupils interpret non-integer answers to division by expressing results in different ways according to the context, including with remainders, as fractions, as decimals or by rounding (for example, $98 \div 4 = \frac{98}{4} = 24 \text{ r } 2 = 24\frac{1}{2} = 24.5 \approx 25$).								•		•		•																								
Pupils use multiplication and division as inverses to support the introduction of ratio in year 6, for example, by multiplying and dividing by powers of 10 in scale drawings or by multiplying and dividing by powers of a 1000 in converting between units such as kilometres and metres.							•	•				•																								
Distributivity can be expressed as $a(b + c) = ab + ac$.																																				
They understand the terms factor, multiple and prime, square and cube numbers and use them to construct equivalence statements (for example, $4 \times 35 = 2 \times 2 \times 35$; $3 \times 270 = 3 \times 3 \times 9 \times 10 = 9^2 \times 10$).	•	•	•				•	•		•	•	•																								
Pupils use and explain the equals sign to indicate equivalence, including in missing number problems (for example, $13 + 24 = 12 + 25$; $33 = 5 \times \square$).																																				

Number – Fractions (including decimals and percentages)	*Stretch and Challenge* Issue 1	2	3	4	5	6	7	8	9	10	11	12	13	14	15	16	17	18	19	20	21	22	23	24	25	26	27	28	29	30	31	32	33	34	35	36
• compare and order fractions whose denominators are all multiples of the same number													•	•																						
• identify, name and write equivalent fractions of a given fraction, represented visually, including tenths and hundredths													•	•				•																		
• recognise mixed numbers and improper fractions and convert from one form to the other and write mathematical statements > 1 as a mixed number [for example, $\frac{2}{5} + \frac{4}{5} = \frac{6}{5} = 1\frac{1}{5}$]														•																						
• add and subtract fractions with the same denominator and denominators that are multiples of the same number														•																						
• multiply proper fractions and mixed numbers by whole numbers, supported by materials and diagrams														•																						
• read and write decimal numbers as fractions [for example, $0.71 = \frac{71}{100}$]															•			•																		
• recognise and use thousandths and relate them to tenths, hundredths and decimal equivalents						•									•			•																		
• round decimals with two decimal places to the nearest whole number and to one decimal place															•																					
• read, write, order and compare numbers with up to three decimal places						•		•							•			•																		
• solve problems involving number up to three decimal places						•		•							•			•																		
• recognise the per cent symbol (%) and understand that per cent relates to 'number of parts per hundred', and write percentages as a fraction with denominator 100, and as a decimal																•	•	•																		
• solve problems which require knowing percentage and decimal equivalents of $\frac{1}{2}$, $\frac{1}{4}$, $\frac{1}{5}$, $\frac{2}{5}$, $\frac{4}{5}$ and those fractions with a denominator of a multiple of 10 or 25																•	•	•																		

	Stretch and Challenge Issue																																			
Notes and guidance (non-statutory)	1	2	3	4	5	6	7	8	9	10	11	12	13	14	15	16	17	18	19	20	21	22	23	24	25	26	27	28	29	30	31	32	33	34	35	36
Pupils should be taught throughout that percentages, decimals and fractions are different ways of expressing proportions.													•	•	•	•	•	•																		
They extend their knowledge of fractions to thousandths and connect to decimals and measures.															•			•																		
Pupils connect equivalent fractions > 1 that simplify to integers with division and other fractions > 1 to division with remainders, using the number line and other models, and hence move from these to improper and mixed fractions.													•	•																						
Pupils connect multiplication by a fraction to using fractions as operators (fractions of), and to division, building on work from previous years. This relates to scaling by simple fractions, including fractions > 1.														•																						
Pupils practise adding and subtracting fractions to become fluent through a variety of increasingly complex problems. They extend their understanding of adding and subtracting fractions to calculations that exceed 1 as a mixed number.														•																						
Pupils continue to practise counting forwards and backwards in simple fractions.																																				
Pupils continue to develop their understanding of fractions as numbers, measures and operators by finding fractions of numbers and quantities.													•	•				•																		
Pupils extend counting from year 4, using decimals and fractions including bridging zero, for example on a number line.															•			•																		
Pupils say, read and write decimal fractions and related tenths, hundredths and thousandths accurately and are confident in checking the reasonableness of their answers to problems.						•		•							•			•																		
They mentally add and subtract tenths, and one-digit whole numbers and tenths.						•									•			•																		

Notes and guidance (non-statutory) Continued	*Stretch and Challenge* Issue 1	2	3	4	5	6	7	8	9	10	11	12	13	14	15	16	17	18	19	20	21	22	23	24	25	26	27	28	29	30	31	32	33	34	35	36
They practise adding and subtracting decimals, including a mix of whole numbers and decimals, decimals with different numbers of decimal places, and complements of 1 (for example, 0.83 + 0.17 = 1).						•									•			•																		
Pupils should go beyond the measurement and money models of decimals, for example, by solving puzzles involving decimals.						•		•							•			•																		
Pupils should make connections between percentages, fractions and decimals (for example, 100% represents a whole quantity and 1% is $\frac{1}{100}$, 50% is $\frac{50}{100}$, 25% is $\frac{25}{100}$) and relate this to finding 'fractions of'.																•	•	•																		
Measurement																																				
• convert between different units of metric measure (for example, kilometre and metre; centimetre and metre; centimetre and millimetre; gram and kilogram; litre and millilitre)																			•	•	•			•	•	•										
• understand and use approximate equivalences between metric units and common imperial units such as inches, pounds and pints																				•	•															
• measure and calculate the perimeter of composite rectilinear shapes in centimetres and metres																									•								•			
• calculate and compare the area of rectangles (including squares), and including using standard units, square centimetres (cm^2) and square metres (m^2) and estimate the area of irregular shapes																								•									•			
• estimate volume [for example, using 1 cm^3 blocks to build cuboids (including cubes)] and capacity [for example, using water]																					•					•										
• solve problems involving converting between units of time																						•				•										
• use all four operations to solve problems involving measure [for example, length, mass, volume, money] using decimal notation, including scaling					•			•	•		•	•	•		•				•	•	•		•	•	•	•										

	Stretch and Challenge Issue																																			
Notes and guidance (non-statutory)	**1**	**2**	**3**	**4**	**5**	**6**	**7**	**8**	**9**	**10**	**11**	**12**	**13**	**14**	**15**	**16**	**17**	**18**	**19**	**20**	**21**	**22**	**23**	**24**	**25**	**26**	**27**	**28**	**29**	**30**	**31**	**32**	**33**	**34**	**35**	**36**
Pupils use their knowledge of place value and multiplication and division to convert between standard units.																			•	•	•	•		•	•	•										
Pupils calculate the perimeter of rectangles and related composite shapes, including using the relations of perimeter or area to find unknown lengths. Missing measures questions such as these can be expressed algebraically, for example $4 + 2b = 20$ for a rectangle of sides 2 cm and b cm and perimeter of 20 cm.																									•								•			
Pupils calculate the area from scale drawings using given measurements.																								•									•			
Pupils use all four operations in problems involving time and money, including conversions (for example, days to weeks, expressing the answer as weeks and days).					•			•	•		•	•	•		•							•				•										

	Stretch and Challenge Issue																																			
	1	2	3	4	5	6	7	8	9	10	11	12	13	14	15	16	17	18	19	20	21	22	23	24	25	26	27	28	29	30	31	32	33	34	35	36
Geometry – Properties of shapes																																				
• identify 3-D shapes, including cubes and other cuboids, from 2-D representations																												•								
• know angles are measured in degrees: estimate and compare acute, obtuse and reflex angles																													•			•	•			
• draw given angles, and measure them in degrees (°)																													•							
• identify:																																				
– angles at a point and one whole turn (total 360°)																																				
– angles at a point on a straight line and $\frac{1}{2}$ a turn (total 180°)																													•			•	•			
– other multiples of 90°																																				
• use the properties of rectangles to deduce related facts and find missing lengths and angles																											•		•							
• distinguish between regular and irregular polygons based on reasoning about equal sides and angles																											•									
Notes and guidance (non-statutory)																																				
Pupils become accurate in drawing lines with a ruler to the nearest millimetre, and measuring with a protractor. They use conventional markings for parallel lines and right angles.																											•		•					•		
Pupils use the term diagonal and make conjectures about the angles formed between sides, and between diagonals and parallel sides, and other properties of quadrilaterals, for example using dynamic geometry ICT tools.																											•		•							
Pupils use angle sum facts and other properties to make deductions about missing angles and relate these to missing number problems.																													•				•			

	Stretch and Challenge Issue																																			
	1	2	3	4	5	6	7	8	9	10	11	12	13	14	15	16	17	18	19	20	21	22	23	24	25	26	27	28	29	30	31	32	33	34	35	36
Geometry – Position and direction																																				
• identify, describe and represent the position of a shape following a reflection or translation, using the appropriate language, and know that the shape has not changed																														•	•	•		•		
Notes and guidance (non-statutory)																																				
Pupils recognise and use reflection and translation in a variety of diagrams, including continuing to use a 2-D grid and coordinates in the first quadrant. Reflection should be in lines that are parallel to the axes.																														•	•	•		•		

Statistics																																				
• solve comparison, sum and difference problems using information presented in a line graph				•																			•												•	•
• complete, read and interpret information in tables, including timetables																																			•	•
Notes and guidance (non-statutory)																																				
Pupils connect their work on coordinates and scales to their interpretation of time graphs.																							•													•
They begin to decide which representations of data are most appropriate and why.																							•												•	•

Cross-curricular links to the National Curriculum Programme of Study

Domain(s)	Topic	Issue	Mathematics: What's the Problem?	Mathematics: Looking for Patterns	Mathematics: The Puzzler	Mathematics: Let's Investigate	English: The Language of Maths	Science: Focus on Science	Computing: Technology Today	Geography: Around the World	History: In the Past / Famous Mathematicians	Art & Design / Music: The Arts Roundup	Design & Technology: Construct	Physical Education: Sports Update	Links to money: Money Matters	Links with home: At Home
Number: – Number and place value	Number	1		•	•	•	•		•	•	•			•		
	Number	2	•	•		•	•				•					
	Number	3	•	•	•	•					•	•				
	Negative numbers	4		•	•	•				•						
Number: – Addition and subtraction	Addition	5		•	•	•	•				•				•	
	Subtraction	6	•	•	•	•	•									
Number: – Multiplication and division	Multiplication	7		•	•	•					•	•		•		
	Division	8	•	•	•	•	•								•	
Number: – Addition and subtraction – Multiplication and division	Mixed operations	9			•	•									•	
	Mixed operations	10		•		•										
	Mixed operations	11			•	•								•	•	•
	Mixed operations	12		•	•			•	•		•			•	•	
Number: – Fractions, (including decimals and percentages)	Fractions	13	•	•	•	•				•		•			•	
	Fractions	14		•	•	•						•				
	Decimals	15		•	•					•					•	
	Percentages	16	•	•		•	•	•				•				
	Percentages	17	•			•		•		•		•	•	•		•
	Fractions, decimals and percentages	18		•	•	•	•									

			Cross-curricular link and *Stretch and Challenge* Feature															
			Mathematics				English	Science	Computing	Geography	History	Art & Design / Music	Design & Technology	Physical Education	Links to money	Links with home
Domain(s)	**Topic**	Issue	What's the Problem?	Looking for Patterns	The Puzzler	Let's Investigate	The Language of Maths	Focus on Science	Technology Today	Around the World	In the Past / Famous Mathematicians	The Arts Roundup	Construct	Sports Update	Money Matters	At Home
Measurement	Length	19	•			•		•		•			•	•		•
	Mass	20	•			•		•			•					
	Volume and capacity	21	•		•	•				•	•		•			•
	Time	22	•			•				•	•					•
	Temperature	23						•	•	•						•
	Area	24	•		•	•		•		•			•			
	Area and perimeter	25	•		•	•								•		
	Measurement	26				•		•		•			•	•	•	
Geometry: – Properties of shapes	2-D shapes	27	•		•	•	•				•		•			
	3-D shapes	28		•	•	•	•						•			
	Movement and angle	29				•							•			
Geometry: – Position and direction	Reflections	30			•	•		•					•			
	Translations	31		•			•			•	•	•	•			
Geometry: – Properties of shapes – Position and direction	Reflective and rotational symmetry	32		•		•	•	•	•	•			•			
	Geometry	33		•		•				•			•			•
	Geometry	34		•	•	•	•			•			•			
Statistics	Statistics	35		•		•	•			•	•	•		•		
	Statistics	36				•	•			•				•		•

Resources used in *Stretch and Challenge 5*

A fundamental skill of mathematics is knowing what resources to use and when it is appropriate to use them. It is for this reason that many of the activities in *Stretch and Challenge* give no indication to the children as to which resources to use. Although the teacher's notes that accompany each activity include a list of resources, children should be encouraged to work out for themselves what they will need to use to successfully complete an activity.

It is assumed that for each activity children will have ready access to pencil and paper, and any other resources that are specifically mentioned in an activity, for example, computers or other Resource sheets. However, all other equipment should be left for the children to locate and use as and when they see is appropriate.

A list of all the resources children are likely to need in *Stretch and Challenge* 5 is given below.

	Stretch and Challenge Issue																	
Resource	**1**	**2**	**3**	**4**	**5**	**6**	**7**	**8**	**9**	**10**	**11**	**12**	**13**	**14**	**15**	**16**	**17**	**18**
pencil and paper	•	•	•	•	•	•	•	•	•	•	•	•	•	•	•	•	•	•
Resource sheets	2, 3	2, 3	2, 3, 19	2, 3	2, 3	2, 3	2, 3	2, 3	2, 3	2, 3	2, 3	2, 3	2, 3, 19	2, 3, 5	2, 3	2, 3, 6, 19	2, 3, 6	2, 3
computer				•		•		•										•
computer with Internet access	•	•	•						•		•	•	•			•	•	
data handling software																•		
ruler				•									•			•		
calculator	•	•		•	•	•	•	•	•	•	•	•		•	•		•	•
concrete apparatus, e.g. interlocking cubes																•		
0–9 digit cards	•							•			•			•			•	
1–16 number cards	•																	
20 different items, each displaying where it was manufactured													•					
0–9 dice															•			
pack of playing cards					•													
set of dominoes					•				•									
A4 paper													•					
counters					•													
material for making a poster, e.g. card, felt-tip pens…						•		•										
lists of one-, two- and three-digit primes, squares and triangular numbers						•		•										
stopwatch							•											
1–6 dice														•				
dozen sized egg carton and 12 counters, or similar														•				
addition (+), subtraction (–), multiplication (×), division (÷), decimal (·), percentage (%) and equals (=) symbol cards											•			•			•	
different breakfast cereals											•						•	
geoboards and elastic bands																		
matchsticks																		
coloured pencils				•														
set of geometric 3-D shapes with a wide selection of different polyhedrons																		
apparatus used for constructing 3-D shapes																		
Centicubes																		
range of different theatre plans – available on the Internet			•															

Resource	*Stretch and Challenge* Issue																	
	19	**20**	**21**	**22**	**23**	**24**	**25**	**26**	**27**	**28**	**29**	**30**	**31**	**32**	**33**	**34**	**35**	**36**
pencil and paper	•	•	•	•	•	•	•	•	•	•	•	•	•	•	•	•	•	•
Resource sheets	2, 3, 19	2, 3	2, 3, 18, 19	2, 3	2, 3, 18	2, 3, 18, 21	2, 3, 18, 19, 21, 22	2, 3, 18	2, 3, 7, 8, 18–21	2, 3, 18, 19, 23	2, 3, 9, 10, 21, 24	2, 3, 8, 11, 18, 19	2, 3, 11, 12, 18, 19, 24	2, 3, 8, 18, 19, 21, 22	2, 3, 11, 18	2, 3, 13, 14, 15, 16, 17, 21	2, 3, 18, 20	2, 3, 18, 20
computer									•		•							
computer with Internet access	•	•	•	•	•	•	•	•		•		•	•	•		•	•	•
data handling software		•			•			•									•	•
ruler	•		•		•	•	•	•	•	•	•	•	•	•	•	•	•	•
calculator	•	•	•	•		•		•				•			•			
coloured pencils				•						•		•		•		•		
material for making a poster, e.g. card, felt-tip pens…									•	•		•		•				
examples of cross-stitching												•	•	•		•		
mirror												•						
ink and / or paints												•						
kaleidoscope														•				
range of colouring materials													•					
poster paper												•	•					
cardboard														•				
scissors			•			•					•		•	•				
camera														•				
different packs of playing cards														•				
two coins, each a different denomination													•					
1–6 dice													•					
street directory and/or map of the local area	•														•			
protractor											•							
A4 paper											•			•		•		
1–8 digit cards																•		
geoboards and elastic bands									•							•		
6 paper squares (per child), ideally each a different colour																•		
interlocking cubes (or similar)							•									•		
map of the world; tape measure; sports cones	•																	
trundle wheel	•						•											
collection of different carrier bags		•																
selection of different daily and weekend newspapers		•						•										•
magazines, food packaging, telephone directory, textbook		•																
selection of items suitable for sending through the post, e.g. book, board game, pair of shoes in a shoe box, packet of pencils or felt tips…		•																
1 litre bucket of water			•															
weighing equipment		•	•															
map of the world showing time zones				•														
analogue clock with geared hands				•				•										
thermometer					•													
weather section from different newspapers					•													
matchsticks						•			•									
Peters Projection map of the world						•												
0–9 dice							•											
stopwatch							•	•										•
boxes; sticky tape; collection of cylinders; different types of small objects such as counters, paper clips, Centicubes, marbles…								•										
simple PE equipment; TV schedules																	•	
collection of six different cereal boxes																		•
set of geometric 3-D shapes with a wide selection of different polyhedrons										•								
apparatus used for constructing 3-D shapes										•								
Centicubes										•								

The Language of Maths

LCM

What is the meaning of the term *Lowest Common Multiple*?

When is it useful to know this?

Illustrate with some examples.

The Puzzler

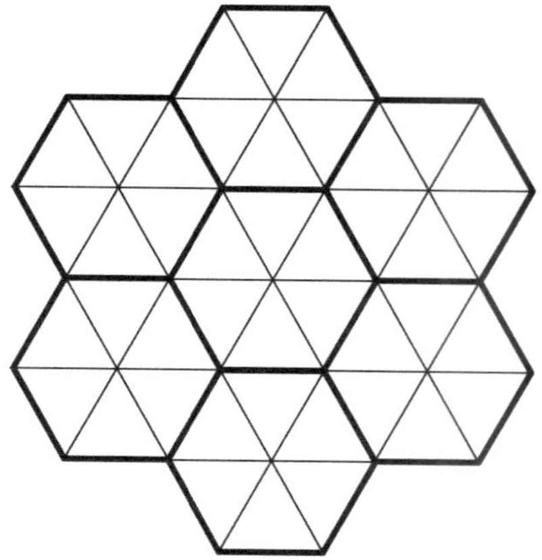

Write each of the digits 1 to 6 in each of the 7 hexagons so that adjacent sides have the same number.

Around the World

The Empire State Building in New York has 1860 stairs. If each flight of stairs has the same number of stairs in it, how many stairs could there be in each flight?

For example, there could be 930 flights, each with 2 stairs in each flight (although this is rather unlikely!).

Investigate how many stairs there are in other tall buildings around the world and work out how many stairs there could be in each flight in these buildings.

S&C Volume 5

The Maths Herald

Name: Date:

Famous Mathematicians

Joseph Lagrange was a French mathematician born in 1736. One of his greatest contributions to mathematics was his work on developing the metric system of measurement.

However, Lagrange was also the first person to prove that every positive whole number can be written as the sum of 4 or fewer square numbers. This proof is referred to as *Lagrange's Four Square Theorem*.

Write down the first 20 square numbers.

Then use these to test *Lagrange's Four Square Theorem* for any number to 400.

Let's Investigate

Investigate the hypothesis that twice the sum of 2 square numbers is always equal to the sum of 2 other square numbers.

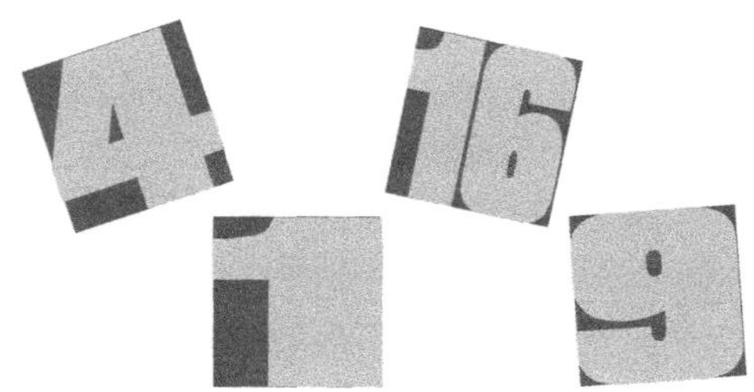

Looking for Patterns

Look at this pattern:

2, 6, 8, 4, 2, 6, 8, 4, 2, 6, 8, 4, 2, 6, …

What is the rule?

What happens to the pattern?

Choose other pairs of 1-digit numbers to start the pattern and apply the rule.

Investigate what happens if you multiply or find the difference.

Looking for Patterns

Look carefully at this diagram.

What's the rule?

What do you notice?

Investigate with other sets of 4 numbers.

What about starting with sets of 2 or 3 numbers?

```
3   7   12   16
  4   5   4   13
    1   1   9   9
      0   8   0   8
        8   8   8   8
          0   0   0   0
```

Looking for Patterns

Make a number sequence. Start by writing any 1-digit number.
Then follow these rules:

- If the number is even, halve it.
- If the number is odd, multiply it by 3 and subtract 1.

Continue the sequence.
What happens?

Choose different 1-digit numbers to start with. What happens?

What if you changed one of the rules?

Technology Today

Investigate how each of the following systems work:

- postcodes
- telephone number area codes
- international telephone number codes
- car registration numbers
- barcodes.

The Puzzler

For this puzzle you will need a set of 1–9 digit cards.

Shuffle the cards and arrange them face up in a 3 x 3 grid formation.

By changing the position of 2 cards at a time, how many moves does it take you to arrange the cards in order smallest to largest?

Once you have completed the puzzle, shuffle the cards and repeat several times. Can you devise some strategies so that it takes you fewer moves to complete the puzzle?

8	3	2
5	6	9
1	4	7

1	2	3
4	5	6
7	8	9

This is considered to be one move.

1	3	2
5	6	9
8	4	7

Sports Update

There are exactly 1000 entrants in a charity run. Each entrant has a number from 1 to 1000. The entrants are given the 1, 2, 3 or 4 digits that are needed to make up their number. These digits are made of fabric and, before the race, entrants have to sew the digits on to a piece of their running kit to make the number they have been given for the race.

So, if your number was 531 you would be given the digits 1, 3 and 5.

How many 1s do the organisers need to have?

How many of each digit 0–9 do the organisers need to have?

How many digits altogether do the organisers need to have?

Let's Investigate

The term *twin primes* refers to 2 primes that have a difference of 2.

For example:

3 and 5

5 and 7.

Write down all the prime numbers to 100.

How many twin primes are there in the numbers 1 to 100? Organise your list of twin primes in ascending order.

For each twin primes work out the total, i.e.

$3 + 5 = 8$

$5 + 7 = 12$.

For each twin primes work out the product, i.e.

$3 \times 5 = 15$

$5 \times 7 = 35$.

What patterns do you notice?

Let's Investigate

Start with 2.

Calculate the answer to 2 to the power of 2.

Calculate the answer to 2 to the power of 3.

Calculate the answer to 2 to the power of 4. When the answer is a 2-digit number (or a 3-digit number or larger) take note of the units digit.

Repeat above several times.

Look at the units digit in each answer.

What has happened?

We have made a 4-link number chain.

2

$2^2 = 2 \times 2 = 4$

$2^3 = 2 \times 2 \times 2 = 8$

$2^4 = 2 \times 2 \times 2 \times 2 = 1\underline{6}$

$2^5 = 2 \times 2 \times 2 \times 2 \times 2 = 3\underline{2}$

$2^6 = 2 \times 2 \times 2 \times 2 \times 2 \times 2 = 6\underline{4}$

$2^7 = 2 \times 2 \times 2 \times 2 \times 2 \times 2 \times 2 = 12\underline{8}$

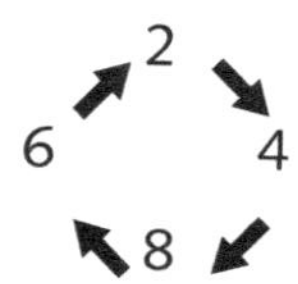

Start with numbers other than 2 and investigate number chains.

S&C Volume 5

The Maths Herald

Name: Date:

In the Past

Louis Braille was born in France in 1809. He invented an alphabet for the blind using raised dots on a 2 × 3 grid.

These are the basic letters and digits of the Braille alphabet.

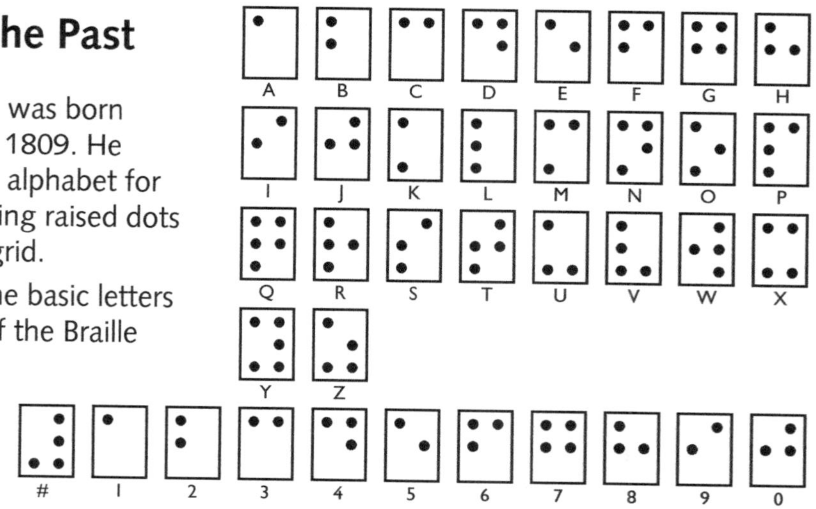

What does this Braille message read?

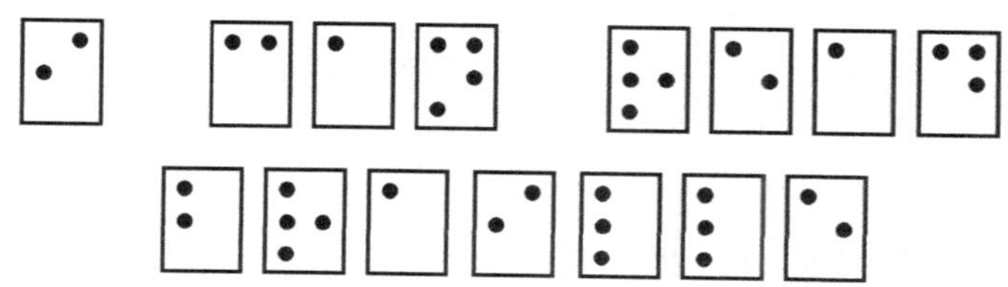

Create a message of your own and give it to a friend to read.

Numbers are made by placing the number sign (#) before the letters a to j. For example:

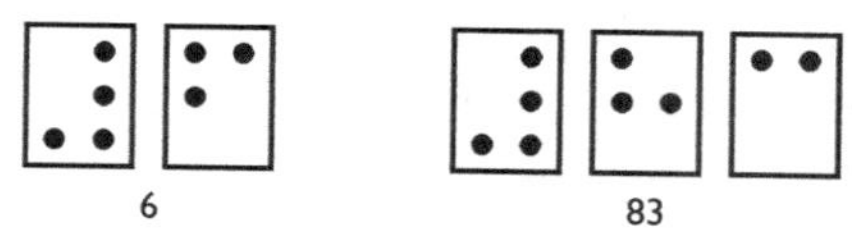

Write some numbers using Braille. Ask a friend to read your numbers.

Looking for Patterns

In Morse code, each letter of the alphabet is represented by a sequence of 'dots' and 'dashes'.

- What are the Morse code sequences for the other letters of the alphabet?
- Investigate the relationship between the number of dots and dashes used for different letters in Morse code and the frequency with which the letters are used.
- As a rule, do the vowels have fewer dots and dashes than the consonants?

A •–
B –•••
C –•–•
D –••
E •

What's the Problem?

There are 10 digits that are used to create numerals in our number system, the Hindu-Arabic system.

- There are 7 different symbols used to represent numbers in Roman numerals. What are they?
- What is the largest Roman numeral you can write using each symbol once only?
- What is the smallest Roman numeral you can write using each symbol once only?

Let's Investigate

Write these numbers written in Roman numerals using the Hindu-Arabic system.

V　X　XXV　XLVI　LXXXIII

Investigate how to write each of these Roman numerals using the Hindu-Arabic system.

$\overline{V}$　$\overline{X}$　$\overline{XXV}$　$\overline{XLVI}$　$\overline{LXXXIII}$

Looking for Patterns

Look at these number sequences. They all have the same rule. What is the rule?

4, 2, 6, 8, 14, 22, 36, 58, …

5, 3, 8, 11, 19, 30, 49, 79, …

2, 7, 9, 16, 25, 41, 66, 107, …

Use the same rule to complete these number sequences.

2, ☐, ☐, 12, ☐, 31, ☐, 81, ☐

☐, 6, ☐, 13, ☐, ☐, 53, ☐, 139

3, ☐, 12, ☐, 33, ☐, 87, ☐, ☐

The Language of Maths

Write down any whole number as a numeral.		63
Write down how many letters there are in the number when it is written as a word.	(sixty-three)	10
Now write down how many letters there are in this number when it is written as a word.	(ten)	3
Keep writing down how many letters there are in the numbers when they are written as words.	(three)	5
What eventually happens?	(five)	4

Try other numbers.

Do you know how to write numbers as words in another language? Does the same thing happen in this language?

What about other languages?

The Puzzler

Find a 3-digit number, so that when the digits of the number are added together the total is equal to the product of the digits.

Can you find more than one 3-digit number?

Find a 3-digit number, so that when the digits of the number are added together the total is one-fifth of the product of the digits.

Can you find more than one 3-digit number?

In the Past

When writing numbers we use the Hindu-Arabic numeral system.

Roman numerals are an example of a different number system.

I	II	III	IV	V	VI	VII	VIII	IX	X	L	C	D	M
1	2	3	4	5	6	7	8	9	10	50	100	500	1000

28 = XXVIII   634 = DCXXXIV   2449 = MMCDXLIX

Investigate different number systems, both past and present.

Write a set of numerals for each system to demonstrate how the number system works.

S&C Volume 5

The Maths Herald

Name: Date:

The Puzzler

Consecutive numbers are numbers that follow each other, e.g. 1, 2 and 3, or 27, 28 and 29.

The proper factors of a number are all its factors except 1 and itself.

75 and 76 are a pair of consecutive numbers that have only four proper factors each. There are 2 other pairs of consecutive numbers less than 100 that also have only 4 proper factors each. What are they?

Let's Investigate

Which prime numbers less than 100 are also prime numbers when their digits are reversed?

Let's Investigate

3 and 5 are prime numbers. They are also consecutive odd numbers.

What other pairs of consecutive odd numbers between 100 and 200 are both prime numbers?

Looking for Patterns

Look at the ones and thousands digits in each of these 4-digit numbers. What do you notice?

What about the tens and hundreds digits?

Each of these numbers is also divisible by 66.

How many other 4-digit numbers are multiples of 66 and have the same ones / thousands and tens / hundreds patterns?

2112
4554
6666

Looking for Patterns

The number 84 has 12 factors (including 1 and itself). These are:

1 and 84
2 and 42
3 and 28
4 and 21
6 and 14
7 and 12

Which other 2-digit numbers have exactly 12 factors?

The Puzzler

Arrange all the digits 1 to 9 to make 4 square numbers, using each of the digits once.

How many different ways can you do it?

What about arranging all 9 digits to make 3 square numbers?

The Arts Roundup

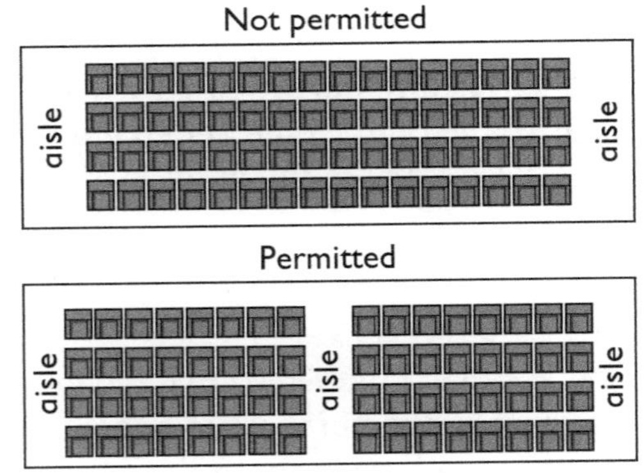

An architectural firm has been asked to design a new performing arts complex consisting of 2 different auditoriums.

The requirements are as follows:

- The Theatre must have a seating capacity of 840 people and must all be on 1 level.
- The Concert Hall must have a seating capacity of 1260 people and must be spread over 2 levels – the stalls and the dress circle.
- All seats must have excellent views of the stages.
- For health and safety reasons, no row can contain more than 10 seats without an aisle. See the example above.

Design a seating layout for the Theatre and Concert Hall. As these are architectural designs they need to be as detailed as possible.

What's the Problem?

1, 2, 3, 4, 6, 7, 8, 9, 12, 14, 16, 18, 21, 24,
28, 36, 48, 56, 63, 72, 84, 112, 126,
144, 168, 252, 336, 504, 1008

The numbers above are all factors of 1008. Are all the factors included in this list or are there any missing? Can you explain a quick way of finding out?

You should be able to use your method to check that you have all the factors for any number.

Around the World

Reykjavik is the capital of Iceland.

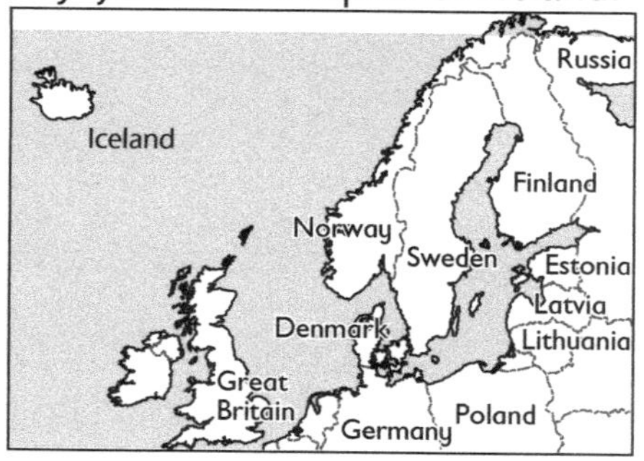

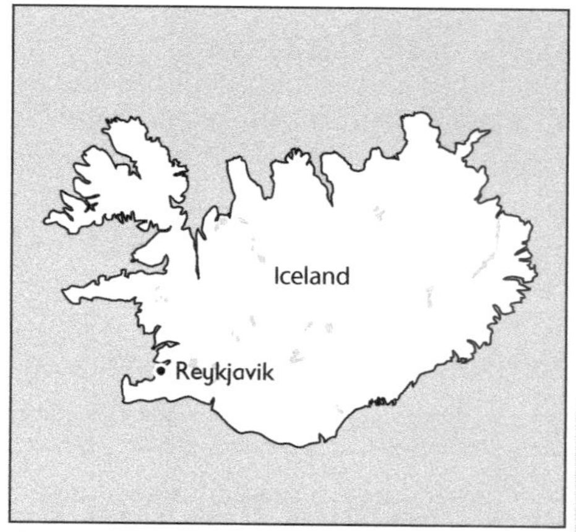

The line graph below shows the maximum and minimum daily temperature averages in Reykjavik.

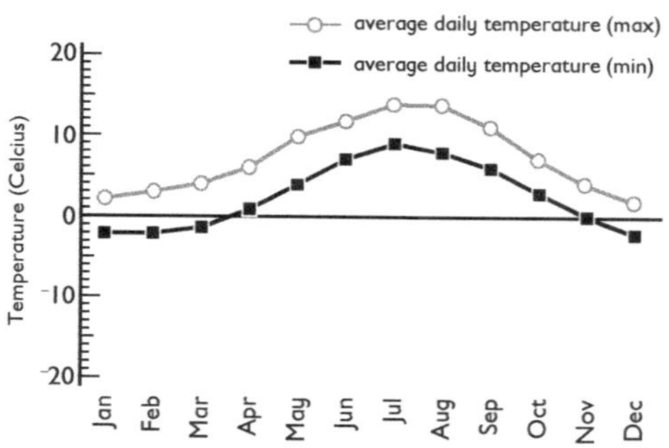

Write a report describing the yearly temperature in Reykjavik.

Include observations regarding the following in your report:

- how the temperature differs between different months of the year
- differences between the average maximum and minimum temperatures.

Let's Investigate

$^-9, ^-8, ^-6, ^-5, ^-4, ^-2, ^-1, 1, 2, 4, 5, 7$

Investigate using 3, 4 or 5 of the numbers above to make zero in as many different ways as you can using only addition.

For example,

$^-6 + 5 + 1$ $\quad$ $^-9 + 4 + 5$

What if you could add and subtract the numbers?

S&C Volume 5

The Maths Herald

Name: Date:

Around the World

These thermometers show the temperatures in six different cities.

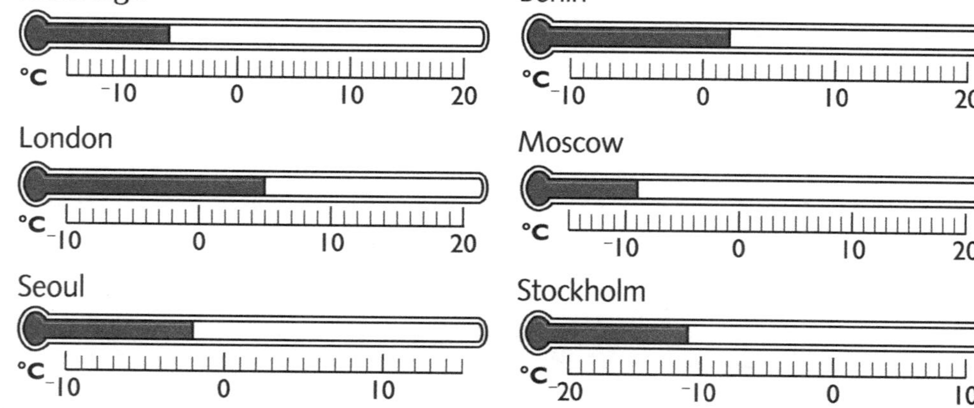

The temperature difference between Seoul and London is 7°C.

Moscow is 2°C warmer than Stockholm.

Choose different pairs of cities and write about their temperature difference.

The Puzzler

In this 'difference wall', subtract the right-hand number from the left-hand number to find the number directly below.

Fill in the empty boxes.

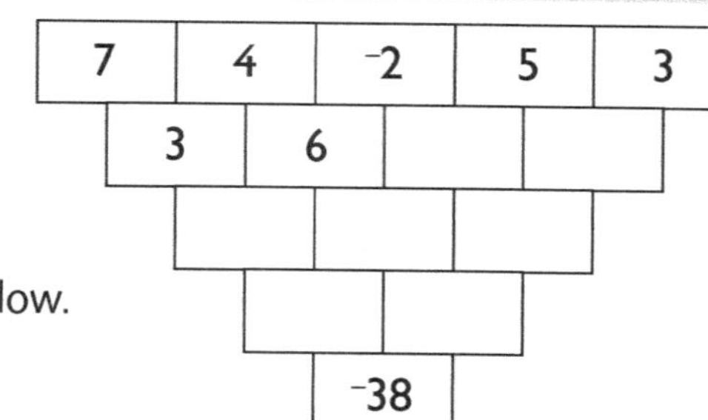

Looking for Patterns

Complete this multiplication table.

×	-5	-4	-3	-2	-1	0	1	2	3	4	5
5											
4											
3											
2											
1											
0											
-1											
-2											
-3											
-4											
-5											

Now, using 3 different colours, shade all the zero answers one colour, all the positive answers another colour, and all the negative answers a third colour.

What patterns do you notice?

What rules can you say about the multiplication of different combinations of positive and negative numbers?

Let's Investigate

Write down pairs of numbers with a difference of 4.

One of the numbers must be a positive number, the other a negative number.

What if both numbers were negative numbers?

Make a difference of 6 / 10 / 15 …

The Puzzler

In a magic square the sum of each column, row and diagonal is the same – this is the magic number.

Complete the magic square.

What is the magic number?

Make up a magic square of your own using a combination of positive and negative integers.

-7	4	-6
	-10	

Around the World

Mount Everest is the world's highest mountain above sea level at 8848 metres (29 029 ft). It is located in the Himalayas.

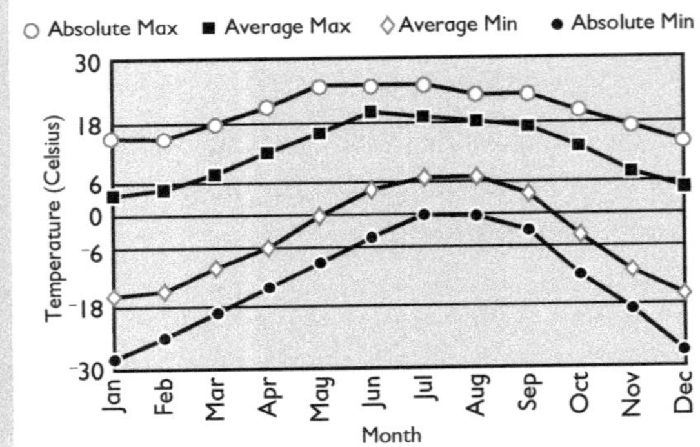

The line graph above shows the average high and low temperatures for each month on Mount Everest. It also shows the maximum and minimum recorded temperatures.

Write a report describing the yearly temperature on Mount Everest.

Include observations regarding the following in your report:

- how the temperature differs between different months of the year
- differences between the average maximum and minimum temperatures
- differences between the maximum and minimum recorded temperatures.

The Puzzler

13 ◇ 12 ◇ 9 ◇ 7 =

Using only the addition and subtraction operations, write a sign in each diamond to make each of the following answers.

23   41   27   3   17

Let's Investigate

If the 3 numbers at the bottom of this addition tower are consecutive and in order, then the top number is 4 times the middle number in the bottom row.

Investigate the above statement.

Write the rule using the symbols *T* and *n*.

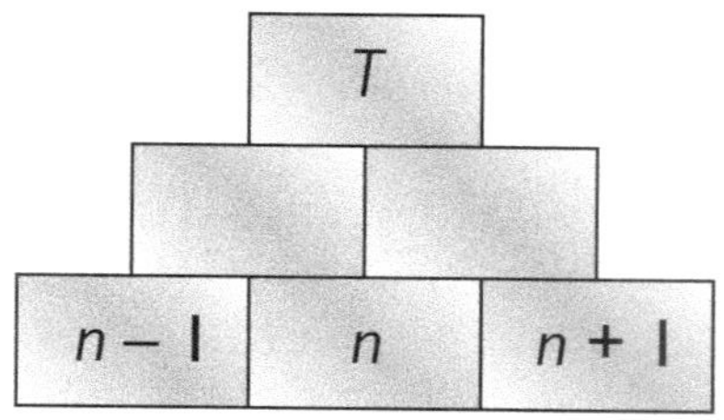

The Puzzler

Using the digits 1–6, investigate writing a number in each circle so that the sum of the 3 numbers along each side of the triangle are the same. Can you do this in different ways?

Write about what you notice about the numbers in the corners of each triangle.

Write down some rules for solving the triangle problem.

What if you used other sets: 0–5, even numbers, a random set?

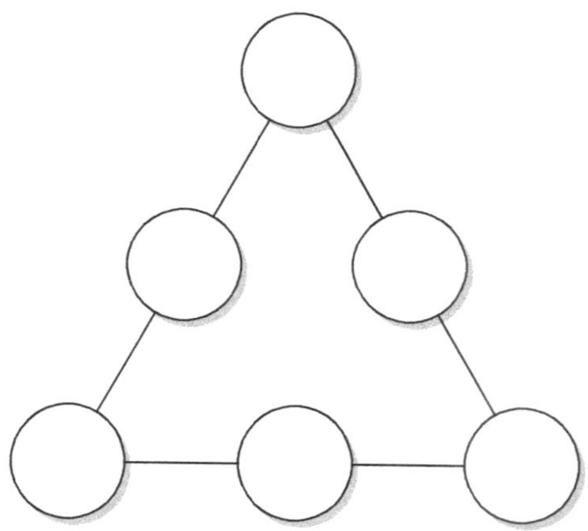

S&C Volume 5

The Maths Herald

Name: Date:

The Puzzler

Look at these 3 counters.

There are numbers written on both sides of each counter. The numbers are all different.

You can only see the numbers written on one side.

The 3 counters are continually spun in the air and each time the 3 numbers on top are added together. It is possible to make all the even totals from 10 to 22 (inclusive), but no odd totals.

What are the numbers written on the other side of each of the counters?

Let's Investigate

The 3-digit number 123 consists of the digits 1, 2 and 3.

The total of these 3 digits is 6 (1 + 2 + 3).

Investigate other 3-digit numbers whose digits add up to 6.

6

Money Matters

You have these coins only.

What different amounts can you make?

What amounts up to £2 can't you make?

The Puzzler

This is a 4 × 4 magic square.

4	9	5	16
15	6	10	3
14	7	11	2
1	12	8	13

Using each of the numbers 1 to 16, the sum of each column, row and diagonal is the same – this is the magic number.

Design your own magic square. Start with a 3 × 3 square, then a 4 × 4 square. What about a 6 × 6 magic square?

In the Past

Remember, in a magic square the sum of each column, row and diagonal is the same – this is the magic number.

Complete these magic squares. Write the missing numbers in Roman numerals.

VIII		
XII		X
XIII		

		IX
	VIII	
VII	XII	

What are the magic numbers?
Write them in Roman numerals.

Try designing your own Roman numerals magic squares.

The Puzzler

Design a 4 × 4 magic square using dominoes.

How many different magic squares can you make?

What about a 3 × 3 magic square made with playing cards?

Record your results.

Looking for Patterns

What is the total of the first 2 odd numbers?
What is the total of the first 3 odd numbers?
What is the total of the first 4 odd numbers?
What is the total of the first 5 odd numbers?
What patterns do you notice?
Can you write a rule?
What is the total of the first 100 odd numbers?

Looking for Patterns

What is the total of the first 2 even numbers?
What is the total of the first 3 even numbers?
What is the total of the first 4 even numbers?
What is the total of the first 5 even numbers?
What patterns do you notice?
Can you write a rule?
What is the total of the first 100 even numbers?

Money Matters

The Happy Eatery are having such a busy night they have got some of the bills and credit card receipts mixed up. It has been a real problem for large tables where the customers have split the bill and 2 or 3 people have paid a proportion of the total bill.

There should be either 2 or 3 credit card receipts for each bill.

Help the Happy Eatery match each bill to the credit cards.

The Language of Maths

Leroy is reading *The Lord of the Rings* by J.R.R. Tolkien. His copy of the book contains 1081 pages.

On how many pages of the book does the number 2 appear in the page numbers?

The Puzzler

In these addition calculations, each number has a difference of 2, 3 or 4 from the numbers in the boxes on either side.

Work out the numbers that belong in each of the boxes.

□ + □ + □ + □ + □ = 100

Difference of: 2 2 2 2

□ + □ + □ + □ + □ + □ = 105

Difference of: 3 3 3 3 3

□ + □ + □ + □ + □ = 120

Difference of: 4 4 4 4

The Language of Maths

Design a poster to show the meaning of these words:

minuend difference subtrahend

The Puzzler

Each of the numbers is either a prime number or a square number.

For each of these subtraction calculations write 3-digit numbers.

□□□ − □□□ = □□□

□□□ − □□□ = □□□

□□□ − □□□ = □□□

S&C Volume 5

The Maths Herald

Name: Date:

What's the Problem?

Last month Malcolm bought a new car.

Today the mileometer shows that he has travelled 1463 miles.

How many more miles does Malcolm have to travel before the digits 1, 3, 4 and 6 next appear on the mileometer?

Let's Investigate

Look at this method of subtraction.

$$\begin{array}{r} 674 \\ -\ 386 \\ \hline \\ \hline \end{array}$$

- Change the minuend to 999 and find the difference.

$$\begin{array}{r} 999 \\ -\ 386 \\ \hline 613 \\ \hline \end{array}$$

- Add the minuend to the difference.

$$\begin{array}{r} 674 \\ +\ 613 \\ \hline 1287 \\ \hline \end{array}$$

- Cross out the most significant digit in the total, and add this to the answer.

$$\begin{array}{r} 674 \\ +\ 613 \\ \hline \not{1}287 \\ +\ 1 \\ \hline 288 \\ \hline \end{array}$$

Choose pairs of 3-digit numbers and use the 999 method to find the difference.

What if you wanted to find the difference between a pair of 4-digit numbers?

What alterations would you need to make to this method?

Find out!

The Puzzler

832 – 65 – 142 + 152

How many different answers can you get for the above calculation by placing brackets in the calculation?

Let's Investigate

Sharon used the digits 1–9 once each to make 3 decimal numbers.

Each number was less than 1 and each number had 3 decimal places.

She wrote them in order:

0·172, 0·493 0·856

What is the difference between Sharon's smallest and largest numbers?

Investigate making 3 decimal numbers so that the difference between the smallest and largest numbers is as small as possible. But remember, you can only use the digits 1–9 once.

What if the difference between the smallest and largest numbers was as large as possible?

Let's Investigate

Look at this diagram.

What do you notice?

What would the next stage of this diagram look like?

Investigate similar diagrams. Start by writing any 4 numbers at the 4 corners of the largest square.

Write about what you find out.

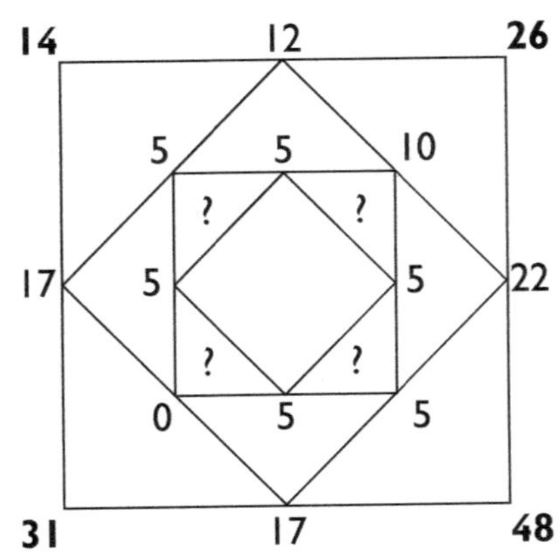

Let's Investigate

5, 6, 7, 8, 9

Cameron used the digits above to make this subtraction with an answer between 1 and 2: 9·5 – 7·68 = 1·82

Investigate using the same 5 digits to make subtraction calculations with answers between 0 and 1.

□·□ – □·□□

Looking for Patterns

Abigail used the digits 1 to 9 to make these calculations.

Work out the answer to each calculation.

Write about the pattern Abigail used.

Work out the answers to the next 2 calculations.

Now find the difference between each answer and the previous one.

Write about any patterns you notice.

12345 – 6789 =
23456 – 7891 =
34567 – 8912 =
45678 – 9123 =
56789 – 1234 =

Let's Investigate

5, 6, 7

Investigate using the numbers above to complete the subtraction below in as many different ways as possible.

9·8 – □·□□

Find the answers.

Arrange the answers in pairs, each with the same total.

Now investigate this subtraction.

9·8□ – □·□

Looking for Patterns

In these 2 calculations each letter represents a different digit.

The same letter represents the same digit in both calculations.

What digit does each of the letters represent?

NOTE:
× is not a digit, it's the multiplication sign.
We've been kind and given you the digit 8.

A T W E × Q = Y E K E 8

```
    8 E
  × W Y
  P W 8   (8E ×  Y)
T S E S   (8E × WS)
T Y S 8
```

Explain how you worked out what each letter represented.

The Arts Roundup

Harriet's ticket for the theatre is in the 7th row from the front, which is also the 17th row from the back.

If there are 17 seats in each row, how many seats are there in the theatre?

Looking for Patterns

Without working out any values, arrange these numbers and calculations in order, smallest to largest.

Then, using pencil and paper or a calculator, calculate the values.

Write another set of 10 numbers or calculations similar to this for a friend to solve. Try and make the difference between the values small so that they are trickier to order!

200×600 $0{\cdot}2 \times 6000$ 6^{10}
10^6 20^3 $2^6 \times 6^2$ $60\,000 \times 200$
$2^6 \times 10$ $2000 \times 0{\cdot}6$ $2 \times 200 \times 2000$

S&C Volume 5

The Maths Herald

Name: Date:

In the Past

In the 1800s, peasants in a remote part of Russia were discovered multiplying numbers using a remarkably unusual method. This method has become known as the 'Russian peasant method of multiplication'.

Look at the example on the right to see how you would calculate the answer to 13 × 18 using this method.

- Halve the numbers in the left column, ignoring any remainders
- Double the numbers in the right column

```
13 × 18
~~6 × 36~~
 3 × 72
 1 × 144
     ---
     234
```

- Cross out all the rows that have an even number in the left column.
- Add the remaining numbers in the right column.

Make up some multiplication calculations of your own. Use the Russian peasant method to work out the answers.

Is this a method of multiplication that you would regularly use? Why? Why not?

Let's Investigate

Write down 3 consecutive numbers.

Multiply together the smallest and the largest numbers.

Square the other number.

Compare the 2 answers. What do you notice?

Choose other sets of 3 consecutive numbers and repeat several times.

What generalisations can you make?

9, 10, 11
$9 \times 11 = 99$
$10^2 = 100$

In the Past

Vedic Mathematics is the name given to the ancient system of Indian mathematics. Try this Vedic Mathematics activity.

Look at this 1–9 multiplication tables square.

×	1	2	3	4	5	6	7	8	9
1	1	2	3	4	5	6	7	8	9
2	2	4	6	8	10	12	14	16	18
3	3	6	9	12	15	18	21	24	27
4	4	8	12	16	20	24	28	32	36
5	5	10	15	20	25	30	35	40	45
6	6	12	18	24	30	36	42	48	54
7	7	14	21	28	35	42	49	56	63
8	8	16	24	32	40	48	56	64	72
9	9	18	27	36	45	54	63	72	81

For each of the 2-digit answers on the square add the digits of each number together. If the answer results in another 2-digit number, repeat until you get a 1-digit number answer. For example:

15: 1 + 5 = 6

48: 4 + 8 = 12 / 1 + 2 = 3   6 and 3 are referred to as the *digital root*.

Using the first 9 × 9 grid on Resource sheet 4, write each digital root in the same position on the square.

Some squares have been completed.

Complete the grid.

1	2	3	4	5	6	7	8	9
2	4	6	8					
3	6	9		6				
4	8							
5		6						
6							3	
7								
8					3			
9								

Now, look at all the 1s on the grid. Join together the centre of each of the squares with a '1' digit in them to the centre of all the other squares with a '1' digit in them.

What do you notice?

Using the other grids on Resource sheet 4, investigate what happens if you do the same for all the squares with a '2' digit in them? What about a '3' or '4' digit in them?

The Puzzler

Investigate the digital roots of cubed numbers.

1 × 1 × 1 = 1 → 1   2 × 2 × 2 = 8 → 8   3 × 3 × 3 = 27 → 9   4 × 4 × 4 = 64 → ?

Is there a pattern?

Investigate for numbers up to 20 × 20 × 20 to check the pattern.

The Puzzler

Write down any 3-digit number. 168

Multiply this number by 7. 1176

Multiply the answer by 11. 12936

Multiply this answer by 13. ☐

What do you notice?

Repeat with other 3-digit numbers.

Can you explain why this happens?

Sports Update

Approximately how many times does your heart beat in a year?

Write about how you calculated your approximation.

How does this compare with the number of times an adult's heart beats in a year?

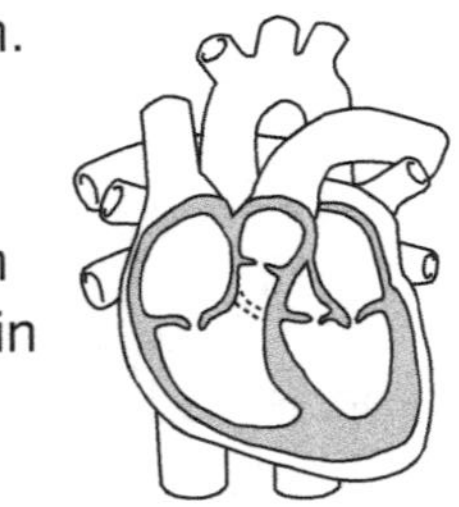

Looking for Patterns

Complete these patterns.

Then write the next 3 calculations in each of these patterns.

Help!
$2^3 = 2 \times 2 \times 2$
$5^3 = 5 \times 5 \times 5$

$1 \times 2 \times 3 \times 4 + 1 = 25 = 5^2$

$2 \times 3 \times 4 \times 5 + 1 = \square = 11^2$

$3 \times 4 \times 5 \times 6 + 1 = \square = \square^2$

$1^3 = 1$

$1^3 + 2^3 = 9 = 3^2$

$1^3 + 2^3 + 3^3 = \square = 6^2$

$1^3 + 2^3 + 3^3 + 4^3 = \square = \square^2$

For each pattern, write down any patterns you notice.

Can you find some similar patterns of your own?

The Puzzler

1 2 3 4 5 6 7 8 9 10 11 12

The product of some of these numbers is 17 280.

What are the numbers?

Write about how you worked out which numbers they were.

Looking for Patterns

One way of finding out whether a larger number is divisible by 11 is to add together the 2 sets of alternate digits.

If both totals are equal or have a difference that is a multiple of 11, then the number is divisible by 11.

For example,

7 1 7 2	1 0 4 1 7	5 3 6 9 3 2
7 + 7 = 14	1 + 4 + 7 = 12	5 + 6 + 3 = 14
1 + 2 = 3	0 + 1 = 1	3 + 9 + 2 = 14
14 – 3 = 11	12 – 1 = 11	

Use each of the digits below only once to complete the 6 numbers so that each number is divisible by 11.

6 0 5 1 4 7 8 9 2 7 1 0 2 0

○ 45 ○   6○ ○27   497○○2

52○ ○08   357 7○○   6○ ○83 8○○

The Puzzler

For each of these division calculations write a 1-digit, 2-digit or 3-digit number.

Each of the numbers is a prime number, a square number or a triangular number.

[ ][ ][ ] ÷ [ ][ ] = [ ][ ]

[ ][ ][ ] ÷ [ ] = [ ][ ]

[ ][ ][ ] ÷ [ ][ ] = [ ][ ]

S&C Volume 5

The Maths Herald

Name: Date:

Money Matters

Michael has a pile of £1 coins.

When he counted the coins into piles of £4, he had £2 left over. When he counted them into piles of £5 he had £1 left over.

How much money might Michael have?

Write some other situations similar to this.

What's the Problem?

There are a total of 440 children eating in the dinner hall today. Mrs Beeton, the school cook, has tables that seat 8 children or 12 children. She knows that with that number of children, she is able to put out just enough tables so that there are no empty seats.

How many tables of each size does Mrs Beeton put out?

Is there more than one possibility?

What patterns do you notice?

What's the Problem?

A school is organising a fun run. They have decided that at the end of the run they will give each child an orange.

They have calculated that they will need a total of 360 oranges.

Oranges come in boxes of 30, 40 or 80.

What are the different combinations of boxes that the school could buy?

The Puzzler

Rearrange each set of 4 digits (and the decimal point) to make a division calculation that equals the number in the box.

Example

· 2 2 3 5

8·4

25·2 ÷ 3

25·2 × 10 = 252

252 ÷ 3 = 84

84 ÷ 10 = 8·4

· 6 4 3 8 → 9·2

· 9 9 3 7 → 5·7

· 6 4 8 6 → 7·8

What's the Problem?

In a large family garage there are a total of 22 wheels.

All the wheels are either on cars or bicycles.

How many cars and bicycles are there in the garage?

How many different combinations of cars and bicycles could there be?

What if the wheels are either on cars, bicycles or tricycles?

Let's Investigate

Draw 2 different factor trees for the number 12.

Investigate and compare different factor trees for multiples of 12 from 12 to 72.

These are 2 factor trees for 20.

20 → 2 × 10 → 2 × 5

20 → 4 × 5 → 2 × 2

The Language of Maths

Design a poster to show the meaning of these words:

divisor dividend quotient remainder

Looking for Patterns

$\square^2 \div 3 = \bigcirc$ r 2

Can you find a square number that, when divided by 3, has a remainder of 2?

Can you predict what remainder you will get when you divide 16^2 by 3?

Let's Investigate

2, 3, 4, 7, 8

Investigate arranging the digits above to complete this division:

□□□·□ ÷ □ =

What is the largest possible answer?

What is the smallest possible answer?

Now investigate this division.

□□·□□ ÷ □ =

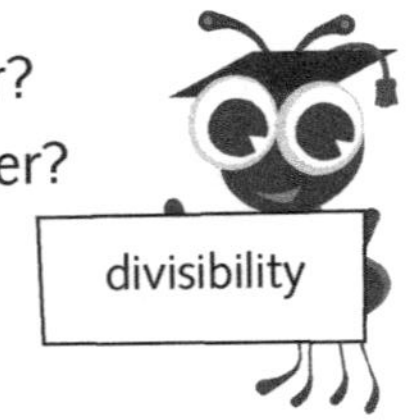

Let's Investigate

Investigate numbers from 1000 to 1200 that are divisible by 6, by 8 and by both 6 and 8.

Show your results on a Venn diagram.

What if the numbers are from 2000 to 2200?

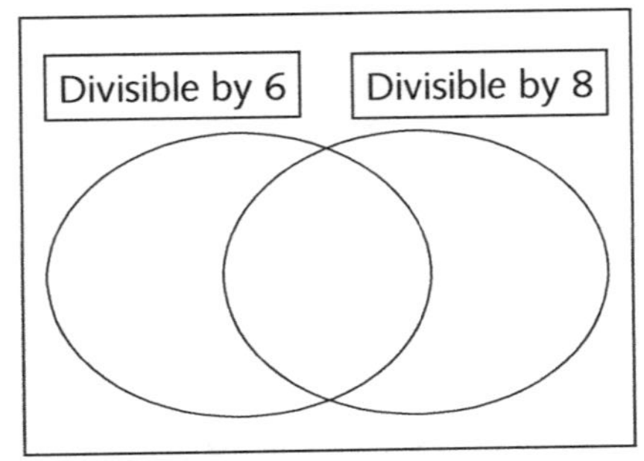

Looking for Patterns

Fill in the missing digits so that each vertical and horizontal number is divisible by 9.

Write about what you did.

	3		■	■	2		4	
■		■	2			■		
■		■		■	3		8	
	7		4	■		■		■
■		■		■				■
■	5	■	6		1		4	

Let's Investigate

Look at this calendar.

S	M	T	W	T	F	S
	1	2	3	4	5	6
7	8	9	10	11	12	13
14	15	16	17	18	19	20
21	22	23	24	25	26	27
28	29	30	31			

A box has been drawn around 4 dates.

Multiply the smallest number by 4 and add 16 to the product. Write down the answer.

Add the 4 numbers together. Write down the answer.

What do you notice about the 2 answers?

Repeat for other sets of 4 numbers.

Does this work for any month?

Can you explain using words or symbols what is happening?

The Puzzler

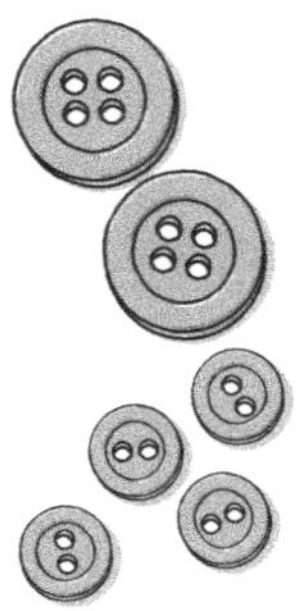

What is the cost of 1 large button, if 2 large buttons and 3 small buttons cost 69p, and 2 large buttons and 4 small buttons cost 80p?

Let's Investigate

Investigate each of the following.

S	M	T	W	T	F	S
	1	2	3	4	5	6
7	8	9	10	11	12	13
14	15	16	17	18	19	20
21	22	23	24	25	26	27
28	29	30	31			

Look at the 4 dates in the 2 × 2 box.

- What do you notice about the sums of pairs of diagonally opposite numbers?
- What do you notice about the difference between the product of pairs of diagonally opposite numbers?

Look at the dates in the shaded column.

- Find the sum of the 5 numbers. What is the relationship between the total and the middle date?

Do these work for any month?

What other amazing patterns are there on a calendar?

Investigate dates in 3 × 2, 3 × 3, 4 × 2 and 4 × 4 boxes.

Investigate 3 dates in a column.

S&C Volume 5

The Maths Herald

Name: Date:

Let's Investigate

How many squares are there on a chessboard?

If you said 64 you would be wrong!

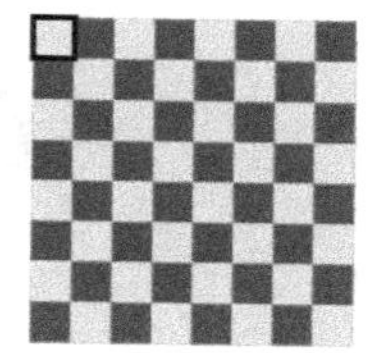

There are a whole lot more, for example:

There is one 8 × 8 square.

There are 16 2 × 2 squares like this (although there are even more 2 × 2 squares).

There are many more different-sized squares on a chessboard.

So, how many different sized squares are there altogether on a chessboard?

Let's Investigate

Step 1: Choose any 3-digit number. 632

Step 2: Multiply the number by 10. 6320 (632 × 10)

Step 3: Find the difference between the numbers in Steps 1 and 2. 5688 (6320 – 632)

Step 4: Divide the answer from Step 3 by the Step 1 number. 9 (5688 ÷ 632)

Repeat several times choosing different 3-digit numbers each time.

What do you notice?

Does this work for 1-digit, 2-digit and 4-digit numbers?

Write an expression or formula to explain why this happens.

Let's Investigate

1 2 3 4 5 6 7 8 9

Arrange the digits 1 to 9 in order, smallest to largest. Combine any number of digits together and insert any of the 4 operations between the numbers.

For example:

12 + 34 – 5 + 67 – 89 = 19

123 × 456 – 789 = 55 299

What's the smallest answer you can make?

What's the largest answer you can make?

The Puzzler

Start with 1000.

Using each of the digits 1 to 9 only once, and any of the 4 operations as many times as you like, write a calculation that equals 0.

1000

0

Let's Investigate

Look at these domino calculations.

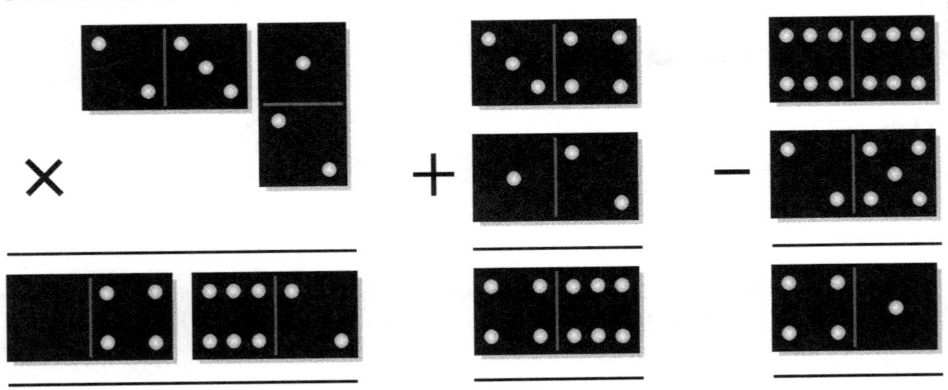

Investigate what other calculations you can make using a set of dominoes.

Money Matters

Do you have a pet?

If so, how much does it cost to keep your pet?

If you don't have a pet, choose a common household pet such as a cat or a dog and investigate how much it costs to keep the pet.

Whether you have a pet or not, calculate how much it costs a month and a year.

Given that the average lifespan is about 12 years for a dog and 15 years for a cat, how much does it cost to keep a dog or cat for its lifetime?

The Puzzler

Look at each of these squares. Fill in the missing numbers.

+ 5 (upwards), + 8 (across)

13			
8	16		
3	11	19	

– 3 (upwards), + 6 (across)

11			

Make some puzzles like these of your own. Think about how you might make them slightly different to the puzzles above.

Let's Investigate

Finally, let's look at how the *casting out nines* method can be used to check the answer to a division calculation.

Look at this example: $567 \div 16 = 35$ r 7

- Cast out nines for the answer (forget about the remainder for the moment). $35 = 3 + 5 = 8$
- Cast out nines for the divisor (the number you are dividing by). $1 + 6 = 7$
- Multiply these 2 numbers together. $8 \times 7 = 56$
- Now add the remainder. $56 + 7 = 63$
- Cast out nines. $6 + 3 = 9$
- Cast out nines for the dividend (the number you are dividing into). $5 + 6 + 7 = 18 \rightarrow 1 + 8 = 9$

$9 = 9$

So the answer is correct.

Choose pairs of numbers and divide the larger number by the smaller number.

Then use the *casting out nines* method to check your answers.

Let's Investigate

- Choose any number less than 40. 15
- Multiply the units digit by 4. 20 (5 × 4)
- Add the tens digit. 21 (20 + 1)

Repeat for the new number.

Look what has happened.

We have made a 6-link number chain.

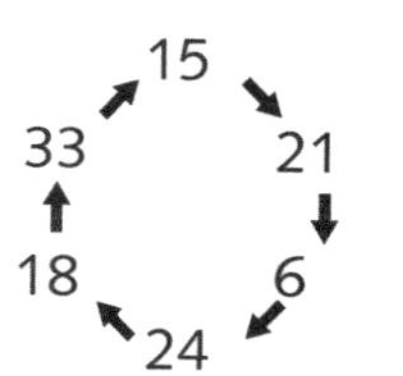

Investigate this rule for other numbers less than 40.

Do they all make chains?

Do the chains all have the same number of links?

What happens if you choose a 2-digit number larger than 40?

What if you changed the rules?

Investigate number chains.

S&C Volume 5

The Maths Herald

Name: Date:

Let's Investigate

The *casting out nines* method is a way of checking the answer to a calculation. It can be used for addition, subtraction, multiplication and division calculations.

Let's begin to see how it works.

A number is divisible by 9 if all the digits in the number add up to 9.

Look at these examples.

477 is divisible by 9 because $4 + 7 + 7 = 18 \rightarrow 1 + 8 = 9$

657 is divisible by 9 because $6 + 5 + 7 = 18 \rightarrow 1 + 8 = 9$

354 is not divisible by 9 because $3 + 5 + 4 = 12 \rightarrow 1 + 2 = 3$

Write down 10 3-digit numbers and see if they are divisible by 9.

Write down 10 3-digit numbers that are divisible by 9.

Let's Investigate

Let's look at how the *casting out nines* method can be used to check the answer to an addition calculation.

The answer to an addition calculation is correct if, after using the *casting out nines* method, the value of the total is equal to the sum of the values of the 2 numbers being added together. Look at these examples.

274 (2 + 7 + 4 = 13 → 1 + 3 = 4)
+ 436 (4 + 3 + 6 = 13 → 1 + 3 = 4)
710 (7 + 1 + 0 = 8)

4 + 4 = 8

8 = 8 So the answer is correct.

538 (5 + 3 + 8 = 16 → 1 + 6 = 7)
+ 275 (2 + 7 + 5 = 14 → 1 + 4 = 5)
803 (8 + 0 + 3 = 11 → 1 + 1 = 2)

7 + 5 = 12 → 1 + 2 = 3

3 ≠ 2 So the answer is incorrect.

Choose pairs of numbers and add them together.

Then use the *casting out nines* method to check your answers.

Let's Investigate

Let's look at how the *casting out nines* method can be used to check the answer to a subtraction calculation.

The answer to a subtraction calculation is correct if, after using the *casting out nines* method, the value of the larger number (the minuend) is equal to the sum of the value of the smaller number (the subtrahend) and the answer (the difference). Look at these examples.

425	minuend
– 168	– subtrahend
257	difference

425 (4 + 2 + 5 = 11 → 1 + 1 = 2)
– 168 (1 + 6 + 8 = 15 → 1 + 5 = 6)
257 (2 + 5 + 7 = 14 → 1 + 4 = 5)

6 + 5 = 11 → 1 + 1 = 2

2 = 2 So the answer is correct.

637 (6 + 3 + 7 = 16 → 1 + 6 = 7)
– 382 (3 + 8 + 2 = 13 → 1 + 3 = 4)
265 (2 + 6 + 5 = 13 → 1 + 3 = 4)

4 + 4 = 8   7 ≠ 8 So the answer is incorrect.

Choose pairs of numbers and find the difference.

Then use the *casting out nines* method to check your answers.

Let's Investigate

Now let's look at how the *casting out nines* method can be used to check the answer to a multiplication calculation.

The answer to a multiplication calculation is correct if, after using the *casting out nines* method, the value of the product is equal to the product of the values of the 2 numbers being multiplied together.

Look at these examples.

53 (5 + 3 = 8)
× 48 (4 + 8 = 12 → 1 + 2 = 3)
2544 (2 + 5 + 4 + 4 = 15 → 1 + 5 = 6)

8 × 3 = 24 → 2 + 4 = 6

6 = 6 So the answer is correct.

76 (7 + 6 = 13 → 1 + 3 = 4)
× 32 (3 + 2 = 5)
2442 (2 + 4 + 4 + 2 = 12 → 1 + 2 = 3)

4 × 5 = 20 → 2 + 0 = 2

2 ≠ 3 So the answer is incorrect.

Choose pairs of numbers and find the product.

Then use the *casting out nines* method to check your answers.

Looking for Patterns

142 857

Multiply the number 142 857 by each of the numbers on a 1–6 dice.

What do you notice?

Sports Update

Brian has a running machine. He uses it every day for a 30-minute workout. However, each day he sets the machine to a different workout programme.

For each day, calculate how long Brian spends on each level of the programme.

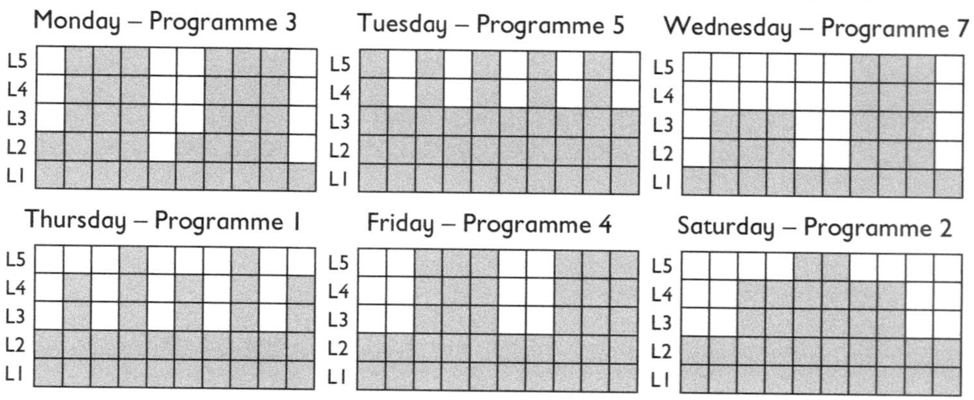

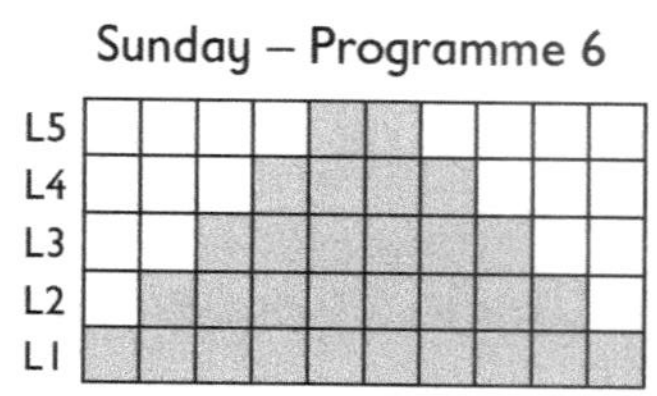

Programme Note
L1 – Easiest Level
L5 – Hardest Level

Rank the days in order of how difficult the workouts are. Start with the day Brian has the easiest workout.

Justify your decisions.

The Puzzler

What is the answer to this calculation?

100 – 99 + 98 – 97 + 96 – 95 + 94 – 93 + 92 – 91 + 90 … + 10 – 9 + 8 – 7 + 6 – 5 + 4 – 3 + 2 – 1 =

Write about how you worked it out.

What about this calculation?

(100 – 99) × (98 – 97) × (96 – 95) × (94 – 93) × (92 – 91) … × (10 – 9) × (8 – 7) × (6 – 5) × (4 – 3) × (2 – 1) =

How did you work out this calculation?

S&C Volume 5

The Maths Herald

Name: Date:

The Puzzler

Look at these calculation mazes. To complete a maze you move from square to square performing calculations as you go until you reach the Exit number.

There are 2 rules to complete a calculation maze.

Rule 1: You must travel through each square in the maze only once.

Rule 2: You must move to a square that is either vertically, horizontally or diagonally adjacent to the square you are currently in.

Complete these calculation mazes. The first one has been started.

Make up some calculation mazes of your own. Have different Enter and Exit locations. Have larger mazes. Perhaps use decimals!

Enter →

6	+	5	+
3	×	4	2
–	3	÷	÷
9	+	5	=

Exit 2

Enter

9	–	+	5
4	+	×	6
×	7	÷	3
8	÷	4	=

Exit 27

Money Matters

Are things more or less expensive now than they were in 1970?

Which items are cheaper now?

Which items are now more expensive?

Can you give possible reasons as to why some things are now cheaper and others more expensive?

Let's Investigate

0 | 1 | 2 | 3 | 4 | 5 | 6 | 7 | 8 | 9

+ | – | × | ÷ | = | · | · | ·

Using some or all of the 0–9 digit cards, any of the operator cards, and the decimal cards, how many different statements can you make involving decimals?

For example:

9 · 1 7 – 8 · 6 4 = 0 · 5 3

Money Matters

The term *cost of living* refers to the cost of a selection of everyday items that most people use.

Compare the increase of this cost to the increase in the average wage in the UK since 1970. Find out:

- on average, how much most people in the UK earn in a week.
- how much most people earned in 1970, per week.
- the cost in 1970 of a range of everyday goods that people buy each week.
- the cost today of the same goods.

How has the cost of living, compared to the average wage, changed since 1970?

At Home

Compare the price and net weight of a range of different breakfast cereals.

Which breakfast cereal offers the best value for money?

Which breakfast cereal offers the worst value for money?

Justify your decision.

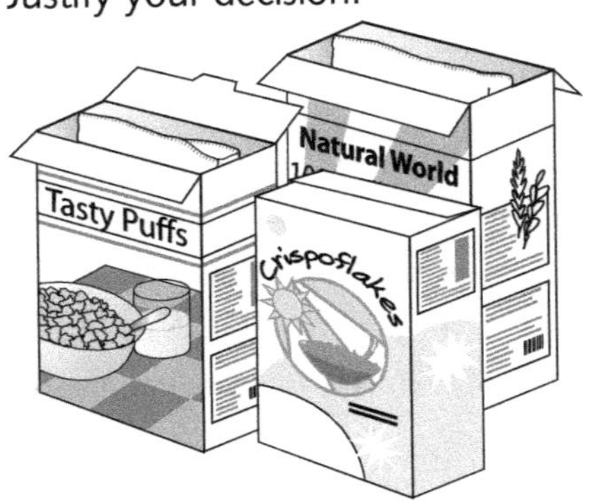

The Puzzler

1 2 3 4

Using all the digits 1 to 4, in ascending order only, can you make each of the numbers from 0 to 30?

Things you're allowed to do!

- Use any of the operations: +, −, x, ÷
- Use fractions
- Use brackets: ( )
- Use the digits to form 2-digit numbers, i.e. 12, 23 or 34

Remember:

- $3^4 = 3 \times 3 \times 3 \times 3$
- $\sqrt{4}$ means the square root of 4 and is equal to 2
- The order of operations

Here are some examples to get you started.

$1 + (2 \times 3) - 4 = 3$ $1 + 2^3 - 4 = 5$ $(1 \times 2) \times (3 + 4) = 14$

Sports Update

In rugby union, a drop goal is worth 3 points. A try with a conversion is worth 7 points.

In a match a team scored a total of 66 points.

All their tries were awarded conversion points and they scored 2 more drop goals than tries.

How many drop goals and converted tries did the team score?

The Puzzler

The sum of my digits is 16.

I am a 3-digit number.

I am a multiple of 2.

The product of my digits is 120.

What number am I?

Make up some of your own 'What number am I?' puzzles.

Try and use a range of different indicators.

Famous Mathematicians

A perfect number is a number that is equal to the sum of its factors other than itself.

The first perfect number is 6 because the factors of 6 are 1, 2, 3 and 6, and 1 + 2 + 3 = 6.

A Greek mathematician called Euclid first described perfect numbers in the 4th Century BC.

There is only one 2-digit perfect number.

Can you work out what it is?

The Puzzler

Nathan Wheeler worked out that the sum of the ages of the 5 people in his family was 111 years.

Nathan also worked out the following:

- The ages of Trevor and Sally total 55 years.
- His age combined with Sally's age total 24 years.
- His age and Helen's age total 50 years.
- Helen's age and Charlie's age total 45 years.

What is the age of each person in the Wheeler family?

Focus on Science

	Boeing 777	Boeing 747	Boeing 737
Maximum number of litres of fuel plane holds	181 280	216 840	26 020
Number of litres of fuel plane uses per hour	7855	15 057	2965
Miles plane travels per hour	590	567	511

Work out the approximate maximum distance that the Boeing 777, Boeing 747 and Boeing 737 can fly on a full tank of fuel.

Be sure to explain how you worked out the answer.

S&C Volume 5

The Maths Herald

Name: Date:

The Puzzler

1	2	3	4

Using each the number cards 1, 2, 3 and 4 only once, what is the largest number you can make my arranging the cards into two 2-digit numbers and multiplying them together?

Sports Update

Joel and Ethan are training for an athletics carnival. Joel takes 6 minutes to run one lap of the running track, and Ethan takes 8 minutes.

How many laps of the running track will Ethan do before meeting up with Joel?

Money Matters

For lunch, Natalie bought a sandwich, a cake and a can of drink for £3.90.

Charlie bought 3 wraps for his lunch. This cost him £3.60.

Louise bought 1 wrap and 2 cakes for her lunch, which cost her £3.

For his lunch Will bought a sandwich, a wrap, and a can of drink. What was the cost of Will's lunch?

Technology Today

Harry worked out these 5 calculations on a calculator but he got each one wrong!

36·5 + 4·78 = *84·3* ✗

234 × 15 = *3645* ✗

6214 – 3506 = *5858* ✗

684 ÷ 9 = *85·5* ✗

4531 + 1875 = *2656* ✗

Explain what Harry did wrong on his calculator.

Focus on Science

An Olympic swimming pool holds approximately $2\frac{1}{2}$ million litres of water.

Each time you flush the toilet it uses around 9 litres of water.

Roughly how many flushes of the toilet would fill an Olympic swimming pool?

Looking for Patterns

07932 462 38

Molly wrote on a scrap of paper the mobile number of a garage she thought might be able to fix her car. However when she went to ring the garage she realised that she had accidentally torn off the last digit of the mobile number.

She remembered thinking when she wrote down the number that it was divisible by 9.

What was the garage's mobile number?

The Puzzler

Be careful!
Don't be confused by X the Roman numeral and x the multiplication symbol.

In this cross-number puzzle clues are given as calculations written in Roman numerals.

Work out each clue and write the answer using Hindu-Arabic numerals.

1 Across has been done for you.

Across

1 CIII – LVI
2 CCCLXXIV + CDLXV
4 III × VIII
6 CXC ÷ V
7 XIX × XXV
8 MMMMCDXVI + MMMCMV
10 MMMMCCXXXVIII – MMDCCCXLIII
12 MMCDXXXI × II
13 MMM + MMMMI
16 CIV ÷ IV
18 M – XLI
20 XVII × L

Down

1 MMDCLI + MDCLXXVII
2 VII × XII
3 DCCCIII – CDXXXII
4 CLXVIII ÷ VIII
5 MMMM + LXXXV
9 CCLX – IV
11 XL × VIII
12 MCMVII + MMCCXXII
14 MMMCCLXXXVI – MDCCXCIII
15 LI × V
17 XVI × IV
19 DCCX ÷ X

What's the Problem?

What number did Ruhl start with?

Explain how you worked it out.

The Puzzler

There are five people in the Lane family.

Their combined age is a total of 120 years.

Mr and Mrs Lane's combined ages is $\frac{2}{3}$ of the total.

2 of the children are twins. The sum of their ages is $\frac{1}{5}$ of the total.

The sum of the children's ages is $\frac{8}{9}$ of Mr Lane's age.

How old is each member of the Lane family?

Write about how you worked them out.

Looking for Patterns

The editor of Global News has a policy regarding what fraction of a page is devoted to news, and what fraction is devoted to advertisements.

Front page: $\frac{9}{10}$ news / $\frac{1}{10}$ advertisements
Pages 2–5: $\frac{4}{5}$ news / $\frac{1}{5}$ advertisements
Pages 6–9: $\frac{7}{10}$ news / $\frac{3}{10}$ advertisements
Pages 10–13: $\frac{3}{5}$ news / $\frac{2}{5}$ advertisements
Pages 14 – penultimate page: $\frac{1}{2}$ news / $\frac{1}{2}$ advertisements
Back page: $\frac{9}{10}$ news / $\frac{1}{10}$ advertisements

Also, the editor will only publish a newspaper in multiples of 18 pages.

If today's edition is 36 pages, what fraction of the newspaper is devoted to news?

S&C Volume 5

The Maths Herald

Name: Date:

Around the World

Find out where 20 different items were produced.

What fraction of them was produced in your local area?

What about elsewhere in Britain?

What about in other parts of Europe?

What about the rest of the world? www

The Arts Roundup

What do these numbers mean?

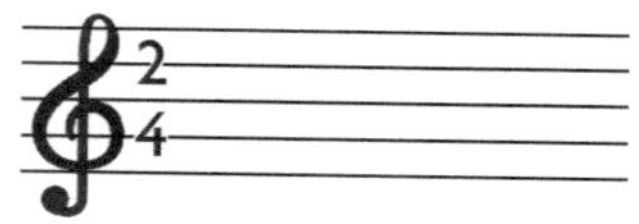

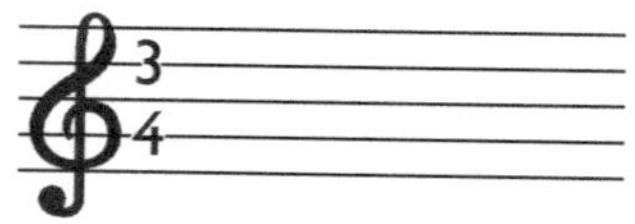

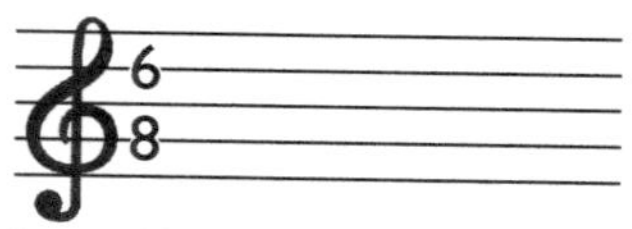

Investigate time signatures in music. www

Let's Investigate

This is a rectangle of 12 squares.

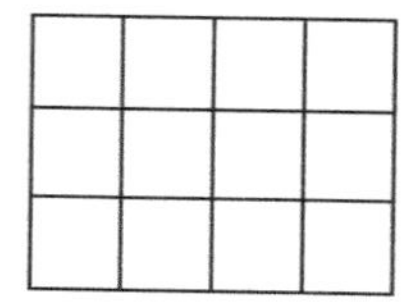

Shade $\frac{1}{2}$ of the squares one colour.

Shade $\frac{1}{4}$ of the squares another colour.

Shade $\frac{1}{6}$ of the squares a third colour.

Shade $\frac{1}{12}$ of the squares a fourth colour.

What has happened?

Can you write this as a fraction calculation?

Using 2 cm squared paper, investigate what other sized rectangles you can draw. Divide them into different unit fractions (fractions with a numerator of 1) and completely shade.

Write each one as a fraction calculation.

The Puzzler

Look at this fraction.

$$\frac{19}{95}$$

If you cross out the 2 digits that are the same you can create an equivalent fraction.

$$\frac{1\not{9}}{\not{9}5} = \frac{1}{5}$$

Find other pairs of fractions where this is true.

Let's Investigate

Find a friend.

Draw a number line. Label it 0, $\frac{1}{2}$ and 1.

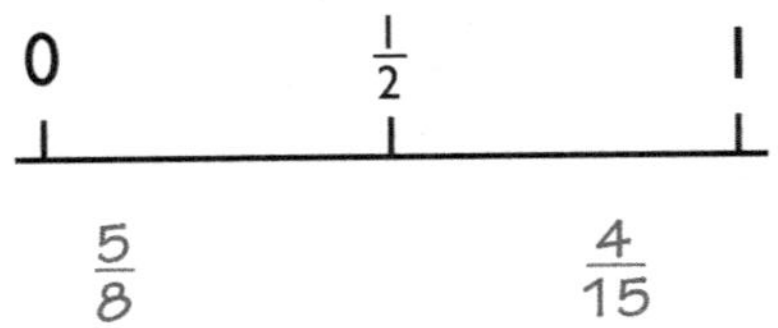

Each person writes a fraction less than 1.

Convert both fractions to a fraction with a common denominator.

$$\frac{5}{8} = \frac{75}{120} \qquad \frac{4}{15} = \frac{32}{120}$$

Now write your fractions on the number line.

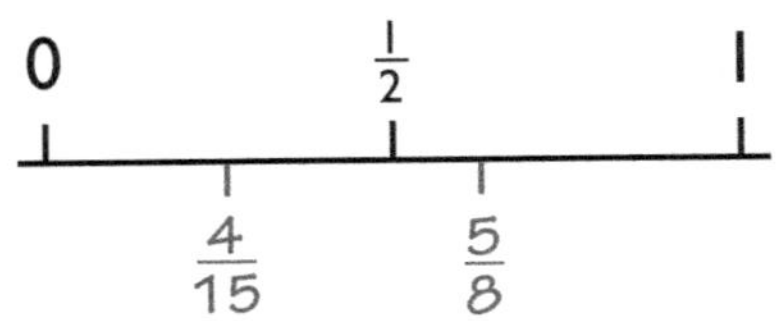

Repeat 5 times.

Discuss together how you could make this into a game.

Write the rules of the game and ask another pair to play it.

The Puzzler

Take a sheet of A4 paper and fold it in half, then in half again. Fold it in half again, and in half once more.

Without opening the paper, how many pieces do you think the sheet has been divided into?

Unfold it and check.

What fraction of the sheet of A4 paper does each small folded piece represent?

Imagine folding the sheet 2 more times than you did at the start of this activity. How many pieces do you think the sheet has been divided into?

What fraction of the sheet of A4 paper does each small folded piece now represent?

Money Matters

The first amount is what fraction of the second amount?

£8 and £72   £6 and £84   £12 and £180

£10 and £390   £9 and £126   £15 and £480

£7 and £84   £5 and £235   £20 and £1740

£13 and £208   £18 and £486   £23 and £529

Let's Investigate

Use only prime numbers from 3 to 19.

The numerator is the sum of 2 of the prime numbers, and the denominator is a different prime number.

Investigate making whole numbers this way.

What pattern do you notice?

For example:

$$\frac{3 + 7}{5} = \frac{10}{5} = 2$$

What's the Problem?

Write a fraction that lies between these two fractions.

$$\frac{1}{2}, \boxed{?}, \frac{1}{3}$$

Let's Investigate

1, 2, 3, 4, 5, 6, 7, 8, 9, 10

Choose 4 of the numbers above to make an equivalent fraction statement, e.g. $\frac{1}{2} = \frac{3}{6}$.

Make as many equivalent fraction statements as you can by choosing 4 different numbers each time.

What if you used 6 different numbers each time from 1 to 20 to make equivalent fraction statements with 3 fractions, e.g. $\frac{1}{4} = \frac{2}{8} = \frac{3}{12}$?

Let's Investigate

Abigail makes 4 different fractions each with 3 as the denominator.

$$\frac{4}{3} \quad \frac{6}{3} \quad \frac{2}{3} \quad \frac{5}{3}$$

She tries to use all the fractions to make 1 by adding or subtracting.

For example: $\frac{6}{3} - \frac{2}{3} + \frac{4}{3} - \frac{5}{3} = 1$

Investigate making 1 by adding or subtracting 4 different fractions, all with a denominator of 3.

What if the 4 fractions had a denominator of 4, 5, 6, …?

Let's Investigate

$$\triangle \div 8 = \square \frac{5}{8}$$

Write a number between 100 and 150 in the triangle.

Complete the division.

Can you find 6 different calculations?

What if the incomplete division was:

$$\triangle \div 8 = \square \frac{7}{8}$$

Let's Investigate

Roll a 1–6 dice 6 times, writing down the number each time.

2, 3, 4, 2, 5, 1

Using 4 or more of the dice numbers, make 2 fractions that when multiplied together give 1, or as near to 1 as possible.

$$\frac{\square}{\square} \times \frac{\square}{\square} = 1$$

For example, $\frac{2}{3} \times \frac{5}{2} = \frac{10}{6} = 1\frac{4}{6}$ or $1\frac{2}{3}$

S&C Volume 5

The Maths Herald

Name: Date:

Let's Investigate

If a 12-egg carton contains 1 egg, it is $\frac{1}{12}$ full.

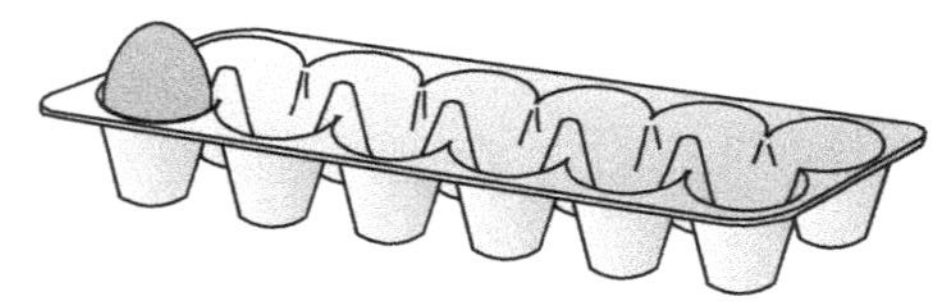

If a 12-egg carton contains 2 eggs, it is $\frac{2}{12}$ (or $\frac{1}{6}$) full.

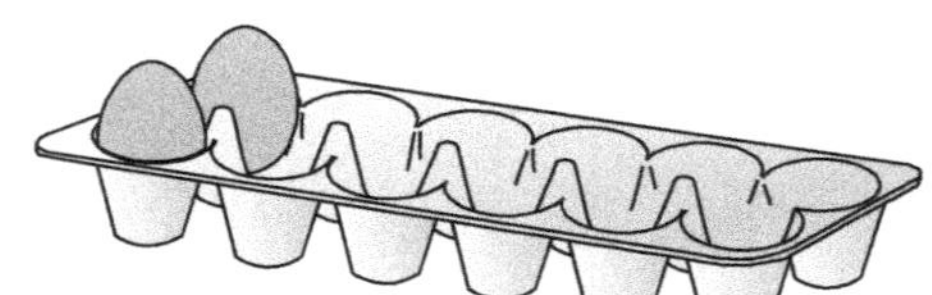

This is a way of seeing how to add fractions. For example, this egg carton shows that $\frac{1}{6} + \frac{1}{3} = \frac{6}{12}$ (or $\frac{1}{2}$).

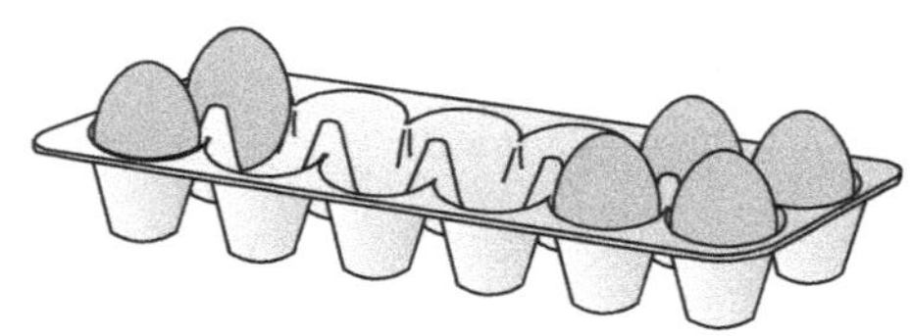

Write different addition fraction calculations for a 12-egg carton that is filled with 1 egg, 2 eggs, 3 eggs, … 12 eggs.

Let's Investigate

This 12-egg carton is $\frac{1}{2}$ full.

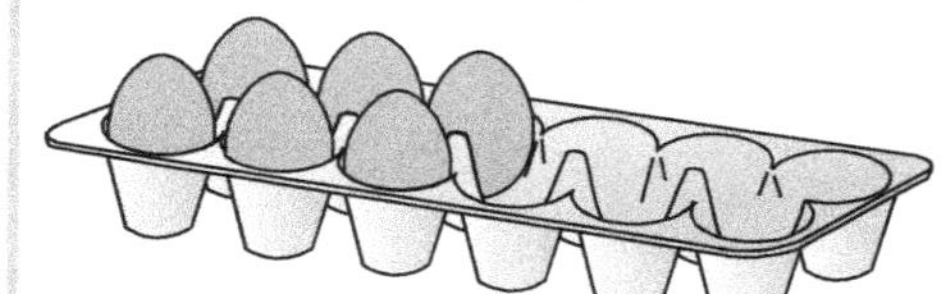

If 4 of the eggs are removed then $\frac{1}{3}$ of $\frac{1}{2}$ is left. This is $\frac{2}{12}$ or $\frac{1}{6}$.

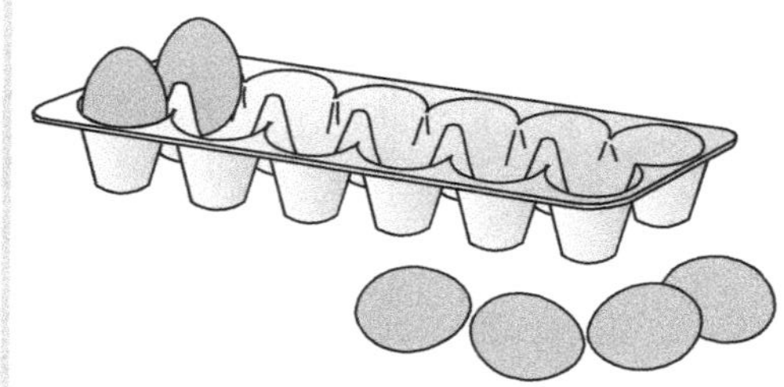

So, $\frac{1}{3} \times \frac{1}{2} = \frac{1}{6}$.

What if you placed eggs in a 12-egg carton so that it was $\frac{1}{3}$ full?

What is $\frac{1}{2}$ of $\frac{1}{3}$?

What do you notice about this answer and the previous answer?

Does this always happen? Investigate.

The Puzzler

The number on each brick is the result of adding the fractions on the 2 bricks below it. Find the missing fractions in these brick walls.

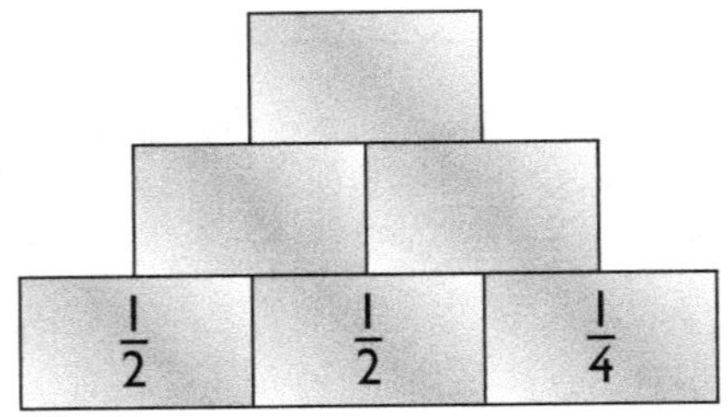

Make up some fraction brick walls of your own, writing all the fractions on the bricks.

Then rewrite your fraction walls leaving out some of the fractions.

Give your fraction walls to a friend to complete.

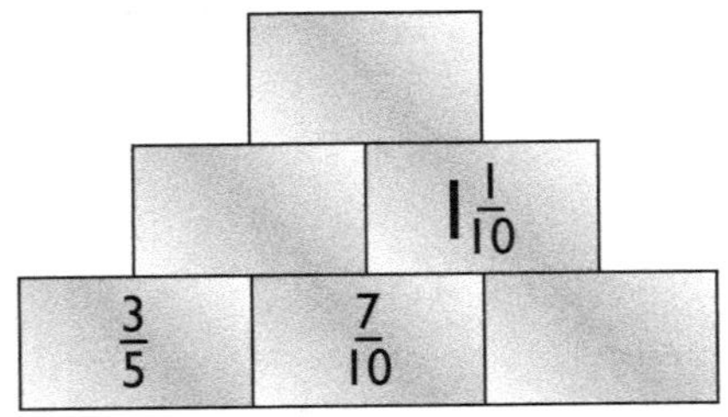

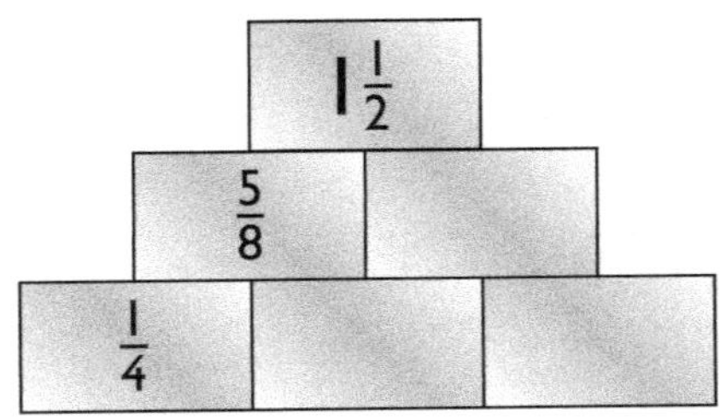

Let's Investigate

0 | 1 | 2 | 3 | 4 | 5 | 6 | 7 | 8 | 9 | × | =

Using some or all of the 0–9 digit cards and the × and = cards, how many different fraction statements can you make?

For example: $\frac{2}{3} \times 6 = 4$

The Arts Roundup

Music is written down using special symbols and notation.

Below are 4 common musical notes, their names and their time values.

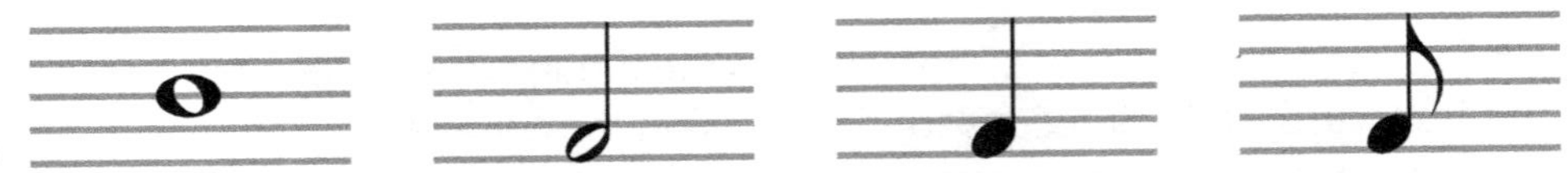

semibreve	minim	crotchet	quaver
1 note	$\frac{1}{2}$ note	$\frac{1}{4}$ note	$\frac{1}{8}$ note

Look at the 2 examples below. They both show one whole note in each bar. The bars are separated by *bar lines*.

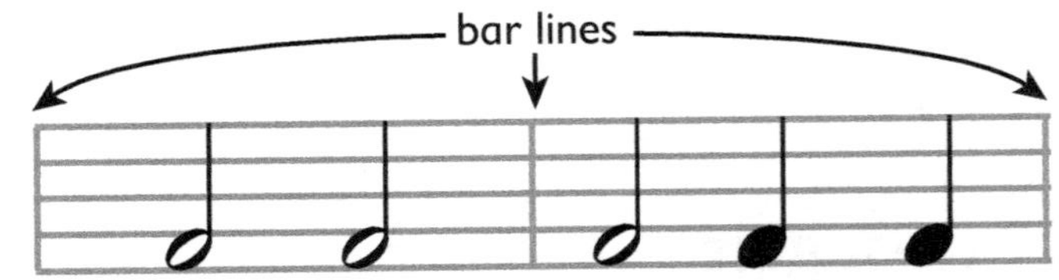

Using Resource sheet 5, write down other ways of composing 1 whole note in a bar.

Looking for Patterns

Look at these addition fraction calculations.

$\frac{1}{2} + \frac{1}{3} = \frac{5}{6}$ $\frac{1}{3} + \frac{1}{4} = \frac{7}{12}$ $\frac{1}{4} + \frac{1}{5} = \frac{9}{20}$ $\frac{1}{5} + \frac{1}{6} = \frac{11}{30}$

What patterns do you notice?

What are the answers to these calculations?

$\frac{1}{6} + \frac{1}{7} = \square$ $\frac{1}{7} + \frac{1}{8} = \square$ $\frac{1}{9} + \frac{1}{10} = \square$ $\frac{1}{10} + \frac{1}{11} = \square$

Money Matters

A 1p coin is worth 0·05 as much as a 20p coin.

A 50p coin is worth 0·05 as much as a £10 note.

- What other coins are worth 0·05 as much as another coin or note?
- What coins or notes are worth 0·2 as much as other coins or notes?
- What about coins or notes that are worth 0·4 as much as other coins or notes?
- What coin is worth 0·0004 as much as another coin or note?

Looking for Patterns

$80 \div 1{\cdot}8 \approx$

If you were asked to approximate the answer to the calculation above you would probably round 1·8 to 2, then divide 80 by 2 to get an answer of 40.

$80 \div 2 = 40$

Is the approximation of 40 more than or less than the actual answer? Why?

S&C Volume 5

The Maths Herald

Name: Date:

Looking for Patterns

Use a calculator to work out the answer to each of these:

$\frac{8{\cdot}4 - 4{\cdot}8}{8{\cdot}4 + 4{\cdot}8}$ $\frac{8{\cdot}4 + 4{\cdot}8}{8{\cdot}4 - 4{\cdot}8}$ $\frac{4{\cdot}2 - 2{\cdot}4}{4{\cdot}2 + 2{\cdot}4}$ $\frac{4{\cdot}2 + 2{\cdot}4}{4{\cdot}2 - 2{\cdot}4}$ $\frac{2{\cdot}1 - 1{\cdot}2}{2{\cdot}1 + 1{\cdot}2}$ $\frac{2{\cdot}1 + 1{\cdot}2}{2{\cdot}1 - 1{\cdot}2}$

What patterns do you notice?

Carry out the calculations above using the numbers 3·1 and 1·3, 6·2 and 2·6, and 9·3 and 3·9 instead.

What do you notice?

Find other similar calculations.

The Puzzler

In a magic square the sum of each column, row and diagonal is the same – this is the magic number.

Complete these magic squares.

What is the magic number for each magic square?

	0·1	
		0·3
		0·4

		3·2
	2·8	
2·4		

Looking for Patterns

Complete the table by finding the missing numerators and denominators as well as the decimal equivalents.

$$\frac{\text{numerator}}{\text{denominator}}$$

		numerator				
				3		8
denominator	2	0·5				
				0·3		
	5		1·2			
						2
	8				0·25	

The Puzzler

Place decimal points in each calculation so that each answer is correct.

34 + 78 = 11·2

78 + 96 + 14 = 31·4

84 – 56 = 2·8

437 + 89 – 264 = 26·2

46 × 128 = 58·88

235 + 612 + 515 = 34·77

55 + 63 + 124 – 307 = 9·97

8456 + 376 + 1482 – 3501 = 54·792

What strategies did you use to work out where the decimal points belong?

Around the World

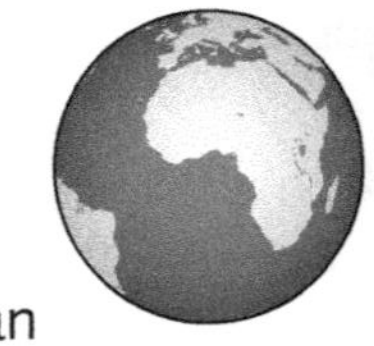

The table on the right shows the land area for each of the seven continents.

By referring to the table we can make the following statements:

The combined land areas of Australia (+ Oceania), Europe and Antarctica are slightly more than the area of Africa.

The land area of Europe is slightly less than one-third of the area of Africa.

Write other statements comparing the land areas of each of the 7 continents.

Continental land areas (in million square kilometres)	
Africa	30·065
Antarctica	13·209
Asia (+ Middle East)	44·579
Australia (+ Oceania)	7·687
North America	24·256
South America	17·819
Europe	9·938

The Puzzler

Using a 0–9 dice, roll the dice 6 times, writing down the number rolled each time.

Use the 6 numbers rolled to complete the 2 calculations.

Place decimal points anywhere in the numbers so that the answers are as close to 50 and 15 as possible.

☐☐ + ☐☐ – ☐☐ = 50

☐☐☐ + ☐☐☐ = 15

The Language of Maths

A *pie chart* is a graph in which a circle is divided into sectors, each representing a proportion of the whole.

The diagram on the left shows how a pie chart is divided into multiples of 10%.

What percentage does each of the sectors of the pie chart on the right represent?

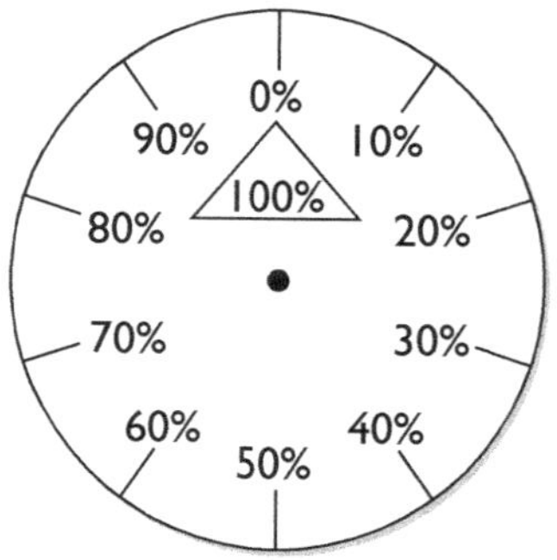

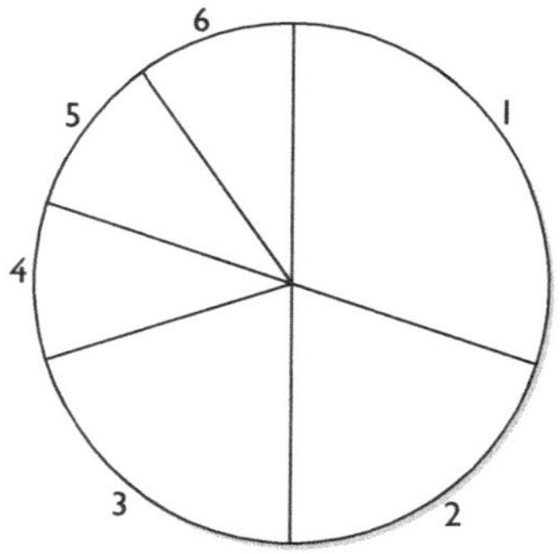

What might this pie chart be about? Give labels for each of the numbers 1 to 6 and a title for the pie chart.

Use one of the multiples of 10% pie charts on Resource sheet 6 to draw a graph of how you spend a typical school day, i.e. in class, travelling, eating, playing, sleeping, doing homework, watching TV and any other activities.

Use other pie charts on Resource sheet 6 to draw graphs for other things, such as what you do over the weekend.

The Language of Maths

The first pie chart has been divided into 100 divisions or percents.

Estimate the percentage of each of the sectors of the second pie chart.

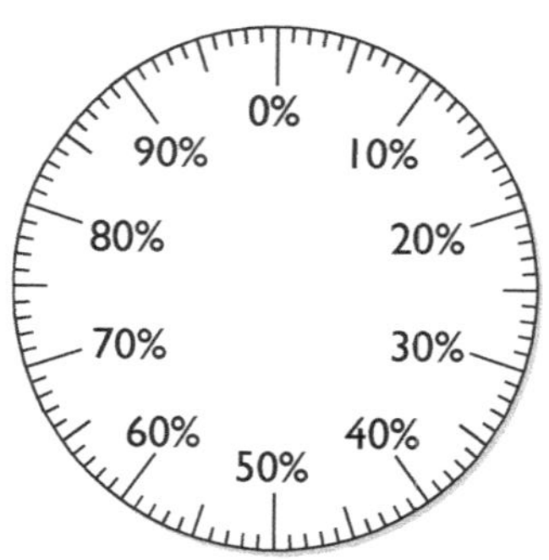

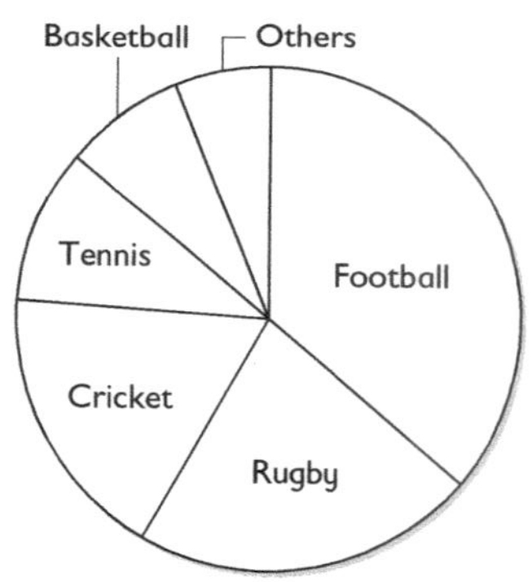

Use other pie charts on Resource sheet 6 to draw graphs for other things, such as the favourite sports of your class.

S&C Volume 5

The Maths Herald

Name: Date:

Focus on Science

There is a large amount of water in fruit and vegetables.

Investigate which fruit and vegetables have the highest and lowest percentages of water in them.

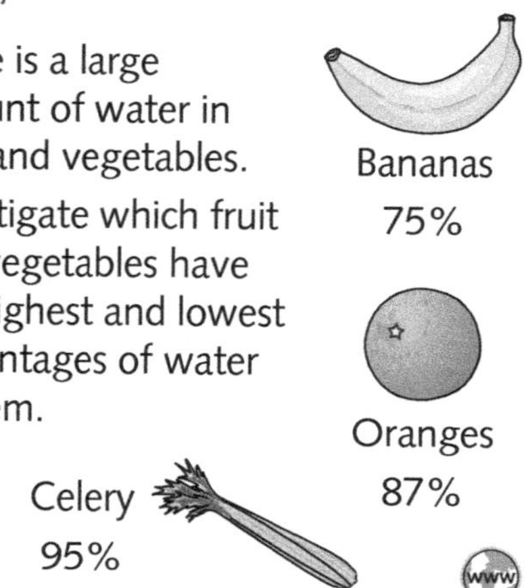

The Language of Maths

What is the meaning of each of these road signs?

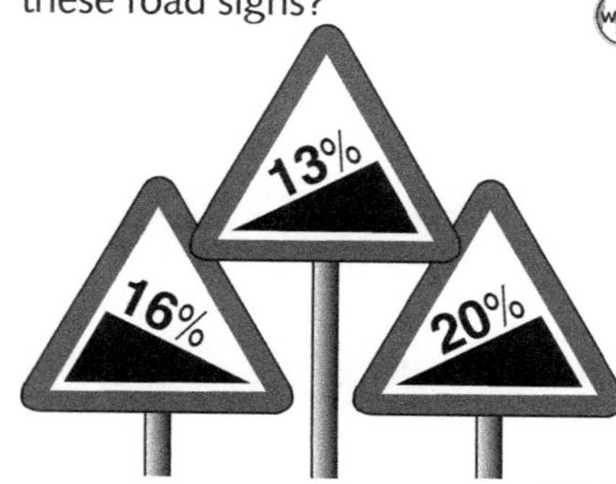

Let's Investigate

50% of a 1–100 number square consists of odd numbers.

10% of the numbers on a 1–100 square are multiples of 10.

90% of the numbers on a 1–100 square are two-digit numbers.

Investigate other percentages on a 1–100 number square. Try and be as creative as you can.

What if you investigated numbers on a 0–99 number square?

1	2	3	4	5	6	7	8	9	10
11	12	13	14	15	16	17	18	19	20
21	22	23	24	25	26	27	28	29	30
31	32	33	34	35	36	37	38	39	40
41	42	43	44	45	46	47	48	49	50
51	52	53	54	55	56	57	58	59	60
61	62	63	64	65	66	67	68	69	70
71	72	73	74	75	76	77	78	79	80
81	82	83	84	85	86	87	88	89	90
91	92	93	94	95	96	97	98	99	100

The Language of Maths

150%

What does 150% mean?

How can you express 150% as a decimal, a fraction and a mixed number?

What's the Problem?

Margot and Jerry Underwood have invited 8 of their closest friends for dinner.

Throughout the evening, Margot is in the kitchen for 40% of the time, while Jerry is alone in the kitchen 10% of the time. They are both in the kitchen together for 20% of the time.

For how much of the evening are either Margot or Jerry alone in the kitchen?

The Language of Maths

Every Friday Mrs Singh gives her class a mental maths test.

This week, she told her class that she wants them to get more than 100% of their normal mark.

Lisa says to Mrs Singh that her words don't make any sense – you can't get more than 100%.

Who is right? Mrs Singh or Lisa? Why?

The Arts Roundup

Smith Studios is staging a new play called *The Pine Tree*. For last week's performances a total of $\frac{7}{8}$ of the tickets were sold.

The Dench Theatre is staging a revival of *Whodunnit*. For last week's performances a total of 86% of the tickets were sold.

Using squared paper, show which theatre sold the greater proportion of their tickets.

Write a statement describing your grids.

Let's Investigate

Using 2 cm squared paper, draw a 2 × 4 rectangle.

How many different ways can you shade 50% of the rectangle?

Here are 2 different ways:

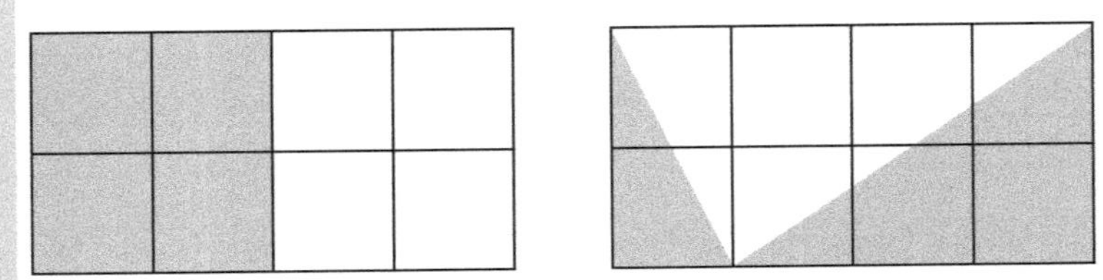

These two are considered to be the same:

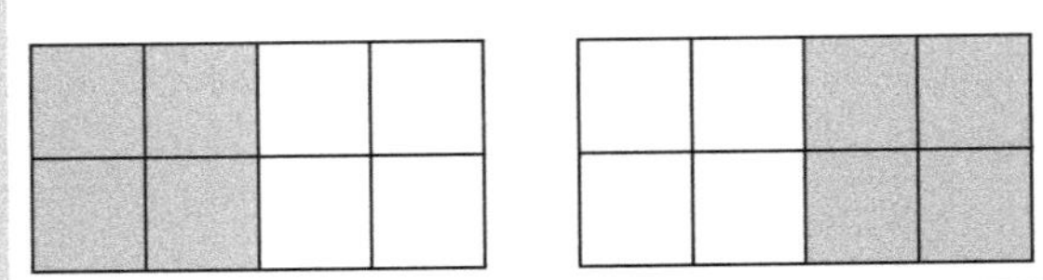

The Arts Roundup

A symphony orchestra is giving a concert. For one of the pieces they are performing they need a total of 106 musicians.

This diagram shows the different sections of the orchestra and what percentage of the total orchestra each section is.

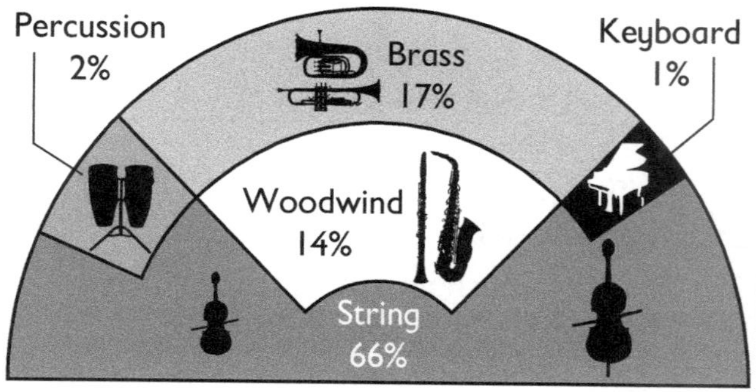

Approximately how many musicians will there be in each section?

The Arts Roundup

A full-sized orchestra (of about 100 musicians) is referred to as either a *symphony orchestra* or a *philharmonic orchestra*.
The United Kingdom has some of the greatest orchestras in the world, such as the London Symphony Orchestra, the London Philharmonic Orchestra, and the Birmingham Symphony Orchestra.

Find out about the size of these or other symphony orchestras.

Does the size of a symphony orchestra vary? Why?

Is the proportion of the different sections of these orchestras similar to those in The Arts Roundup activity above?

Smaller-sized orchestras are often referred to as *chamber orchestras*.

Approximately how many musicians are there in a chamber orchestra?

Find out if the proportions of the different sections of a chamber orchestra are the same as for a symphony orchestra. www

S&C Volume 5

The Maths Herald

Name: Date:

Let's Investigate

0	1	2	3	4
5	6	7	8	9

% × =

per

Using some or all of the 0–9 digit cards and the %, × and = cards, how many different statements can you make?

For example: 4 0 % × 9 5 = 3 8

Focus on Science

An adult has, on average, 5·5 litres of blood pumping through their body.

Using the pie chart, calculate approximately how many litres / millilitres this is for each part of the body.

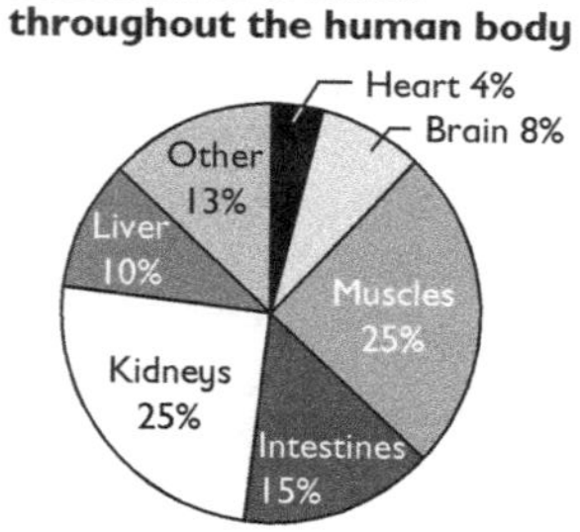

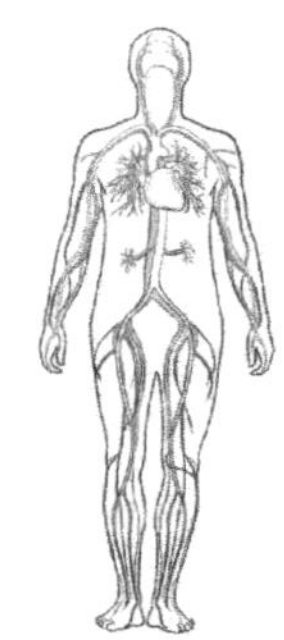

At Home

Look at the list of ingredients on a range of different breakfast cereals.

Write down what percentages of protein, carbohydrate and fat the different cereals have.

Using the net weight printed on the cereal box, calculate how many grams of each ingredient the different cereals have.

Which of the cereals is the healthiest?

Which is the unhealthiest?

Explain your reasoning.

Around the World

William's Junior School is trying to raise £1000 throughout the year for a charity that builds schools in Africa. The aim is for each year group to raise £250.

Look at this table. It shows what proportion of their target each year group has achieved so far.

	Autumn Term		Spring Term		Summer Term	
	1st half	2nd half	1st half	2nd half	1st half	2nd half
Year 3	$\frac{1}{5}$	$\frac{3}{10}$	$\frac{1}{10}$			
Year 4	0·2	0·15	0·05			
Year 5	25%	25%	20%			
Year 6	10%	10%	10%			

Do you think that William's Junior School will reach their target?

Justify your conclusions.

What's the Problem?

Natural World is a health food store. They make their own brand of muesli.

This is their recipe.

They sell their muesli in small (500 g), medium (1 kg) and large (2 kg) bags.

Calculate how many grams of each ingredient there are in a small, medium and large bag of Natural World's 100% All Natural Fruit and Nut Muesli.

Natural World's
100% All Natural
Fruit and Nut Muesli

35% wholemeal oat flakes
20% wholemeal wheat flakes
15% wholemeal corn flakes
10% dried and chopped dates, banana, apple and apricot
5% dried and chopped fig, plum, pear and peach
5% sultanas
2% linseed
2% sunflower seeds
2% halved and toasted pecans
2% chopped hazelnuts
2% chopped almonds

Sports Update

Every day Tim does some exercise.

The amount of time varies each day, but the routine is always the same.

The pie chart shows the proportion of time Tim spends on each exercise activity as part of his daily routine.

Calculate how much time Tim spends each day on each of the 4 different exercise activities.

Day	Time spent exercising
Monday	30 minutes
Tuesday	1 hour
Wednesday	30 minutes
Thursday	1 hour
Friday	30 minutes
Saturday	2 hours
Sunday	$1\frac{1}{2}$ hours

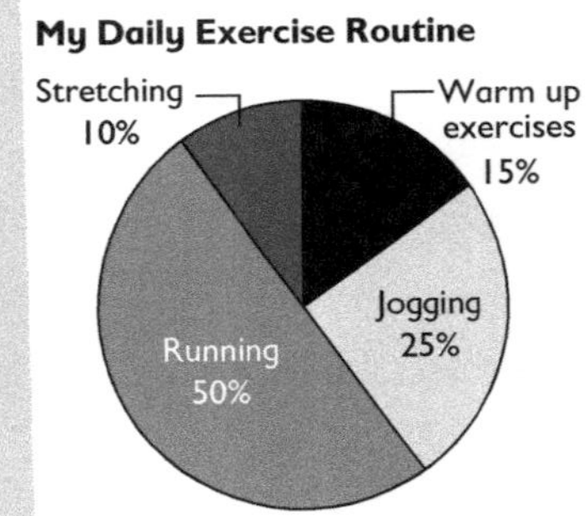

How many minutes does Tim spend each week on the 4 different exercise activities?

Write about how you worked out the answer.

Construct

What percentage of the walls in your classroom are windows?

Is this the same for all the classrooms in your school?

Which classrooms have the greatest and the least percentage of windows?

Is this the same for the dinner hall / gym?

What are the benefits of having a high percentage of wall area compared to windows? What are the disadvantages?

The Puzzler

Match the fraction, decimal and percentage that together total 1.

$\frac{9}{25}$	$\frac{2}{5}$	$\frac{23}{100}$
$\frac{7}{50}$	$\frac{3}{10}$	
$\frac{6}{25}$	$\frac{7}{20}$	$\frac{1}{20}$

0·6, 0·73, 0·42, 0·21, 0·12, 0·28, 0·38, 0·55

15%, 18%, 3%, 67%, 43%, 17%, 48%, 53%

Make a problem like this where the fraction, decimal and percentage together total 2.

Let's Investigate

Remember, to convert a fraction to a decimal we divide the top number, the *numerator*, by the bottom number, the *denominator*.

For example,

$\frac{1}{2} = 1 \div 2 = 0{\cdot}5$

$\frac{1}{4} = 1 \div 4 = 0{\cdot}25$

Make different lists of fractions that when converted to decimals have:

- 1 decimal place
- 2 decimal places
- 3 decimal places
- recurring digits.

Look at the lists you have made.

Write some rules so that a friend will know which list a given fraction will belong to, without having to work it out.

Now show your rules to a friend and ask them to apply them to some of the fractions in your lists. (But don't show your friend which lists the fractions are in!)

Does your friend put them into the same lists as you?

If not, work with your friend to find out why not.

S&C Volume 5

The Maths Herald

Name: Date:

The Puzzler

Join together 3 or more adjacent squares vertically, horizontally and / or diagonally that total 1.

How many different ways can you find to do this?

0·148	0·104	0·225	0·15	$\frac{5}{8}$
$\frac{1}{4}$	$\frac{1}{5}$	0·41	0·177	0·151
0·393	0·36	0·03	0·236	0·13
0·292	$\frac{2}{5}$	0·275	0·069	$\frac{3}{8}$

The Language of Maths

Harriet found that $\frac{1}{7} = 0{\cdot}142857142857142\ldots$

What do the 3 dots at the end of the digits mean?

We can write $\frac{1}{7}$ as $0{\cdot}\dot{1}4285\dot{7}$

What do you think the 2 dots mean?

Investigate the recurring decimal fractions for $\frac{2}{7}, \frac{3}{7}, \frac{4}{7}, \frac{5}{7}$ and $\frac{6}{7}$.

What if the fractions were $\frac{1}{13}, \frac{2}{13}, \frac{3}{13}, \frac{4}{13}, \ldots \frac{12}{13}$?

Looking for Patterns

Have you ever wondered what a fraction looks like as a pattern?

Probably not. But you're about to find out!

Remember, to convert a fraction to a decimal, we divide the top number, the *numerator*, by the bottom number, the *denominator*.

For example,

$\frac{1}{2} = 1 \div 2 = 0{\cdot}5$

$\frac{1}{4} = 1 \div 4 = 0{\cdot}25$

Sometimes it helps to use a calculator to convert a fraction to a decimal.

For example,

$\frac{2}{7}$ = [2] [÷] [7] [=] 0·285714285714 …

$\frac{7}{11}$ = [7] [÷] [1] [1] [=] 0·63636363636 …

What do you notice about the decimals in the last 2 examples above?

Here is a way of representing these decimals as patterns.

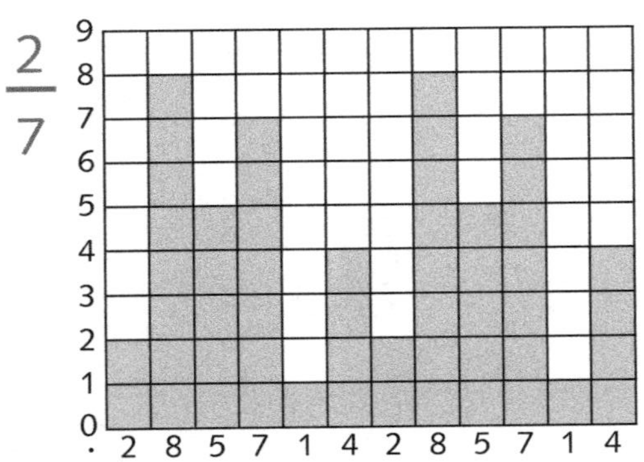

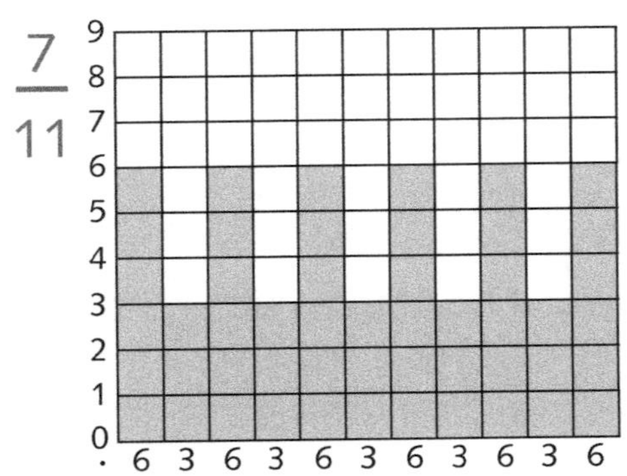

What fractions make interesting decimal patterns? Investigate.

What does your date of birth look like as a pattern?

For example, 14th June 2001: 1406200114062001 …

What about today's date?

What other dates make interesting patterns?

What would the time on a 24-hour digital clock look like as a pattern?

For example, 16:34 would be 1634163416341634 …

What about other times?

How could you use 2 different colours to show the addition of 2 numbers as a pattern?

For example, 3541 + 5038 = 8579.

Investigate.

The Puzzler

0·6 | 75%	0·8 | 10%	1 | 50%	40% | $\frac{3}{4}$
$\frac{1}{2}$ | 5%	$\frac{1}{100}$ | 0·4	100% | $\frac{1}{5}$	20% | 0·05
$\frac{1}{10}$ | 25%	$\frac{4}{5}$ | 30%	1% | $\frac{3}{5}$	0·3 | $\frac{1}{4}$

- Arrange the 12 dominoes above into the 3 arrangements below. All touching sides of the dominoes must be equivalent.

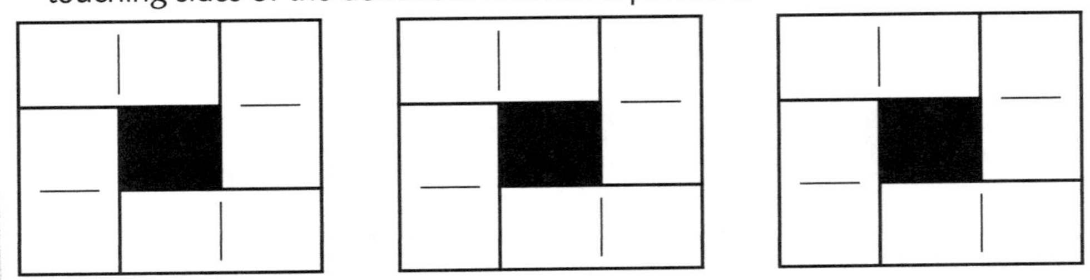

Looking for Patterns

Write down 20 percentages from 1% to 100%.

For example 37%.

Imagine that each of the percentages represents the amount shaded in a 100 square.

As a decimal, write what fraction of each 100 square is not shaded.

What do you notice?

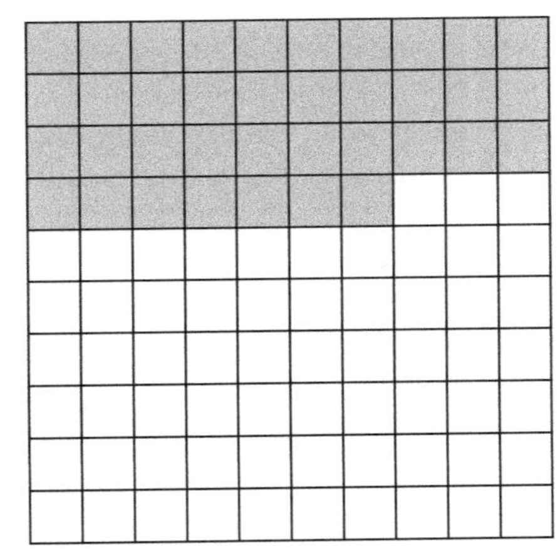

What's the Problem?

Paul and Derek have a meeting together in Leeds.

Paul lives 60 kilometres closer to Leeds than Derek.

If they both leave at the same time and Paul travels at 80 kilometres an hour and Derek travels at 90 kilometres an hour, they will both arrive at the same time.

How long will it take them both to reach Leeds?

Sports Update

Measure the length of your step. If you could take 1 million steps, how many kilometres would you travel?

How many steps would it take you to travel 20 km?

At Home

Find some photos of yourself.

Can you estimate the height of a tree, a building or another person in the photo by comparing it with what your approximate height was in the photo?

Construct

Using estimations only, lay out on the ground 4 sports cones so that they form the vertices of a square, each side approximately 10 metres in length.

Now measure each side and diagonal of the square.

How accurate was your square?

S&C Volume 5

The Maths Herald

Name: Date:

Around the World

Using a map of your local area, compare the distances between different landmarks.

Write at least 5 sentences. Try and be as accurate as possible with your comparisons.

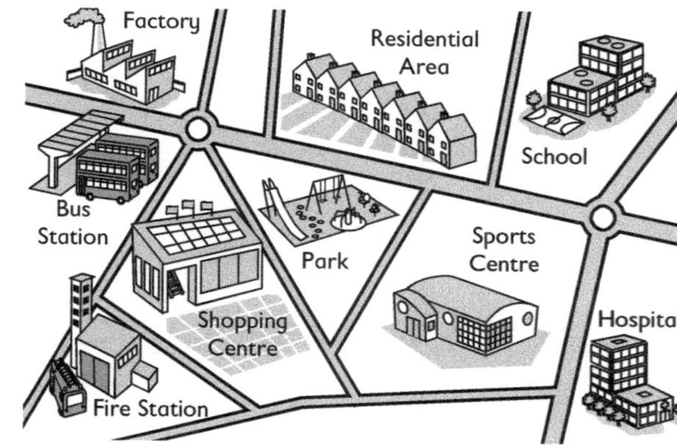

Let's Investigate

Using squared paper, mark 6 different points on the grid lines. Include some points, such as Points B and E, which are half way between the intersection of a pair of vertical and horizontal lines.

Find the shortest distance between different points following the grid lines.

For example, the distance between Points D and E is 4·5 units.

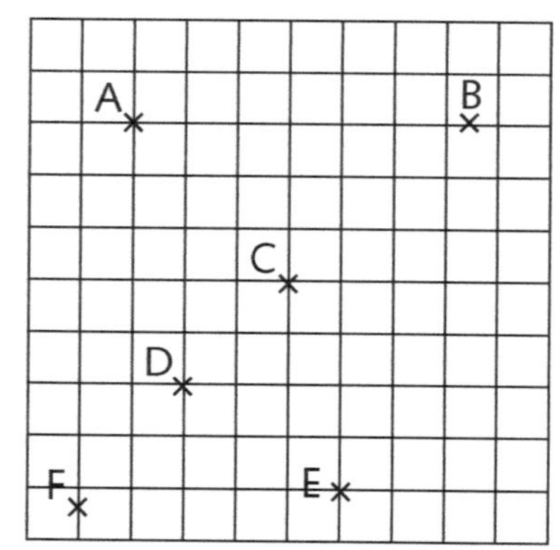

What's the Problem?

How far is 1 million metres?

Can you name 2 cities in the world that are approximately 1 million metres apart?

How far is 1 million kilometres?

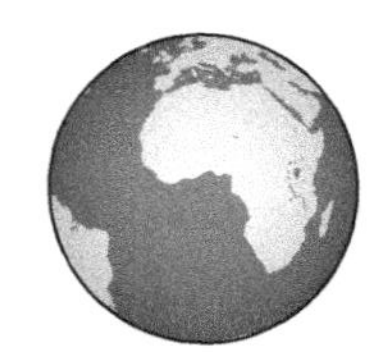

Let's Investigate

How thick is this sheet of paper?

How thick is the material in the clothes you are wearing?

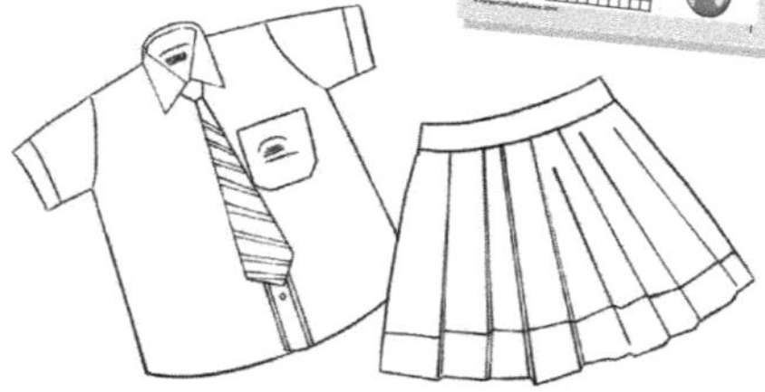

The answers to these questions are obviously very small measurements.

You probably don't have access to a measuring instrument accurate enough to be able to make these measurements.

Work out a way to calculate these measurements.

When you have thought of a method, test it out.

Then write about your method, including its degree of accuracy.

Around the World

Around the World in 80 Days is a famous book written by a Frenchman called Jules Verne.

Find out the different types of transport the characters used in the book to travel around the world in 80 days.

Can you trace the voyage on a map of the world? Include some references as to how long it took to get between places.

What's the Problem?

Last Monday, David drove from his office to Glasgow for a meeting on Tuesday morning.

He calculated that he travelled the 455 miles at an average speed of 65 miles per hour.

If David arrived at his hotel in Glasgow at 6:25 pm, at what time did he leave his office?

What's the Problem?

Ruby gave her grandson Michael an 8-volume set of books she had stored in a box in the attic.

When he got home Michael noticed that a moth had eaten its way through the back cover of Volume 1 through to the last page of Volume 8.

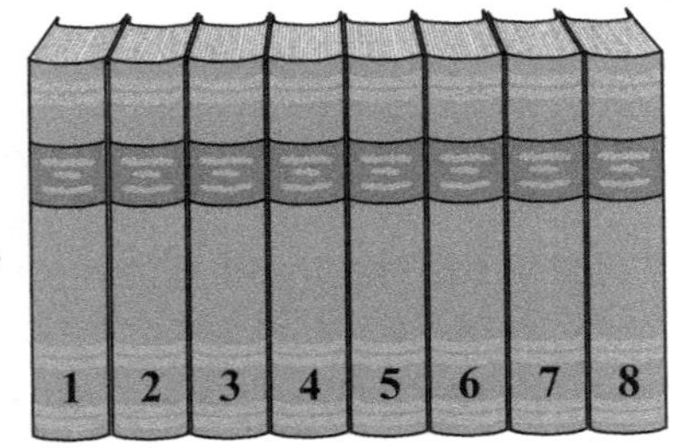

The book covers are $\frac{1}{2}$ cm thick and the pages of each volume are 5 cm thick.

What is the total number of centimetres that the moth ate?

What's the Problem?

Gillian drove to Bristol to visit her mother. It took her a total of 8 hours to drive from her home to her mother's home.

On her way back home, there was much less traffic, and Gillian was able to travel on average 5 km per hour faster. It took Gillian half an hour less to travel back home.

- What was Gillian's average speed on the way to her mother's?

- What was Gillian's average speed on her way back home?

Focus on Science

Gloria is a keen gardener. But she is troubled by the caterpillars that love to eat her climbing roses.

Gloria has a climbing rose that is attached to a trellis that is 2 metres tall.

She has noticed that there is one particular caterpillar that each day manages to climb up 10 cm. However, at night because of the dew, it slips back 6 cm. How many days will it be before the caterpillar reaches the top of the trellis?

Let's Investigate

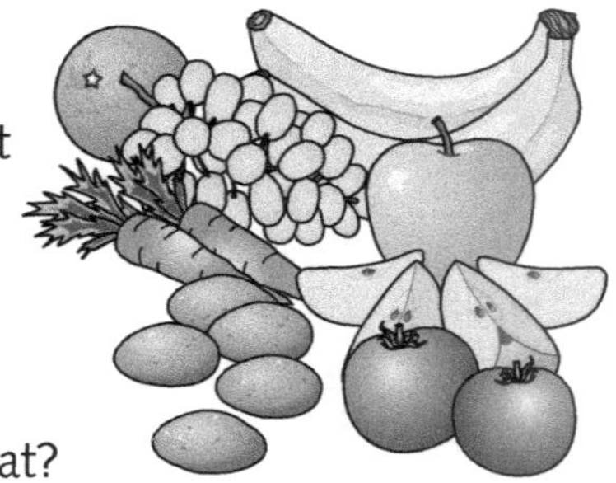

Health recommendations are that we eat at least 5 portions of fruit and vegetables each day.

Investigate the mass of different fruits and vegetables.

If you eat 5 portions of fruit and vegetables in a day, what mass of fruit and vegetables will you eat?

What about in a week?

What about in a month?

Let's Investigate

Investigate the cost of sending a selection of parcels through the post.

Gather together a selection of suitable parcels, e.g. book, board game, pair of shoes in a shoe box, packet of pencils or felt tips.

Weigh each parcel in kilograms and grams and measure each parcel in centimetres and millimetres.

Record all of the information in a table.

Using the Royal Mail 'Price Finder' and 'Weight Comparison' tables, find out how much it would cost to post each of these parcels to an address in the UK.

How would the cost of sending each parcels change if it was posted to an address abroad?

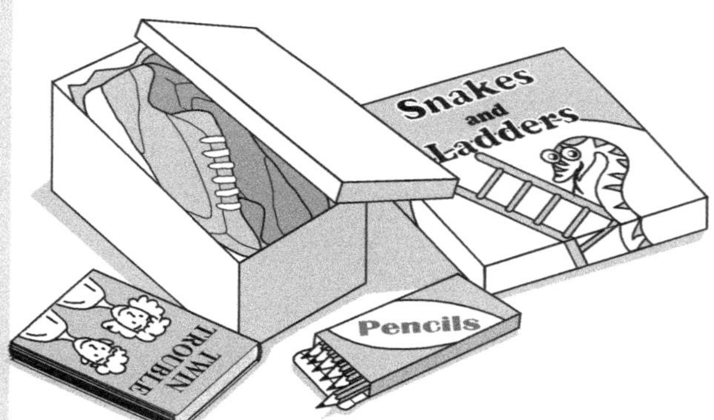

S&C Volume 5

The Maths Herald

Name: Date:

Focus on Science

Who has the strongest carrier bag?

Gather together a collection of carrier bags from different supermarkets and other shops.

Devise a test to find out how strong each bag is.

Write about your test and how effective it is as a measure of strength.

Record your results in a table, chart or graph.

What's the Problem?

3 oranges and 2 lemons have the same mass as 4 oranges and 1 clementine.

1 orange has the same mass as 1 lemon and 2 clementines.

How many clementines have the same mass as 1 orange?

What's the Problem?

May and June are the 2 busiest months at The William's Strawberry Farm.

For these 2 months of the year, workers at the farm work 7 days a week.

On average, a total of 145 kilograms of strawberries are picked each day. However, for every kilogram of strawberries that are picked, 100 grams cannot be sold as they are bruised or damaged in some way.

Altogether how many kilograms of strawberries is The William's Strawberry Farm able to sell in May and June?

What's the Problem?

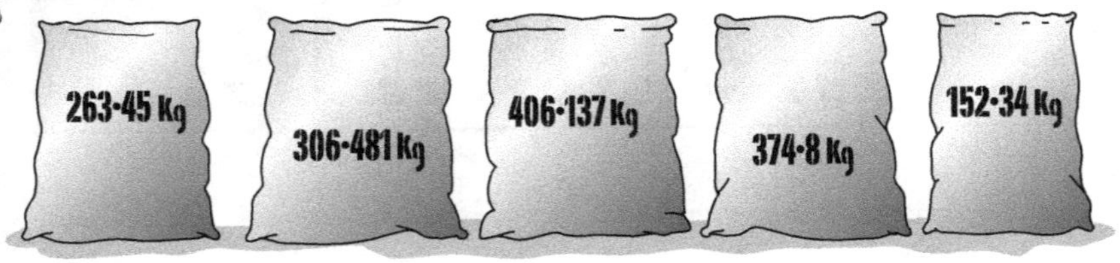

Choose 2 or more of the above weights to make each of the following amounts:

Let's Investigate

What is the mass of this sheet of paper?

How are you going to find out?

Investigate the mass of different types of paper and cardboard. For example:

- newspapers
- magazines
- food packaging
- telephone directory
- cover of a textbook
- pages inside a textbook.

In the Past

Many ancient civilisations had a weight system that used units known as the *mina*, *shekel* and *talent*. However, the values of these units varied between different civilisations.

Find out about the Babylonian, Sumerian and Roman mina, shekel and talent.

In the Past

In 1826, the *imperial system* of weights and measures was introduced in the UK. This is still the system used in many places today. However, many of the units of weight are no longer in common usage.

Investigate these imperial units for mass and weight and their equivalents.

For example: 14 pounds = 1 stone

- dram
- pound
- stone
- hundredweight
- ton
- ounce
- clove
- tod
- sack
- quarter

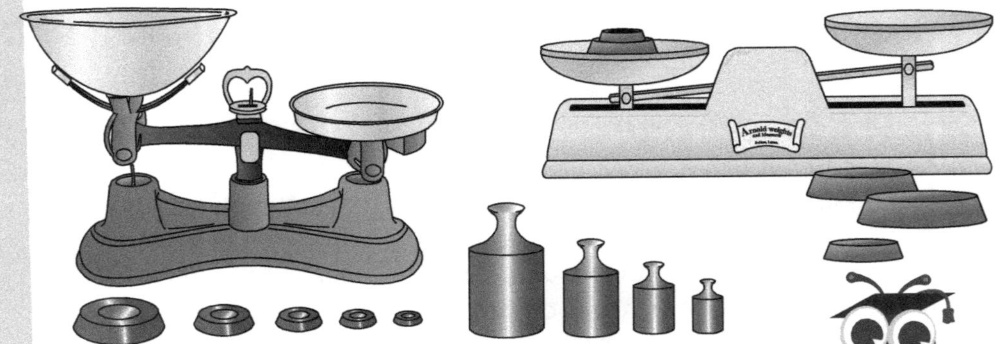

In the Past

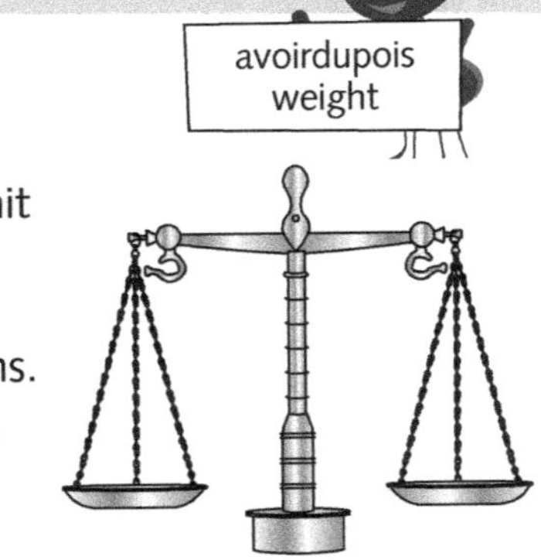

A seed of wheat or barleycorn was the earliest unit of mass. This seed was known as a *grain*. Larger units were referred to as a *stone*. The *pound* was derived from the *mina* used by ancient civilisations.

While the *mina* is no longer used, we still use the *stone* and *pound* and even the *grain* as standard units of weight.

Until the imperial system was introduced in England in 1826, ounces and pounds had different values. For the horseman, the stone weighed 14 pounds. For wool the stone was 16 pounds, while the butcher and fishmonger used a stone of only 8 pounds.

Investigate the history of the pound and the stone.

In the Past

In the 18th century the British introduced a standardised way of measuring volume and capacity.

The *gallon* was the basic English unit of volume and was originally the volume of 8 pounds of wheat. This standard unit led to other units for different sizes such as the *quart, pint, peck* and *bushel*.

Larger volumes were carried in *barrels*, *hogsheads* and other containers. Their sizes in gallons varied and couldn't always be accurately assessed.

In 1826 the imperial system was introduced, and in this system a gallon was designed to hold exactly 10 pounds of water.

Investigate the following imperial units for volume and capacity and their equivalents. For example:

1 quart = 2 pints
4 quarts = 1 gallon

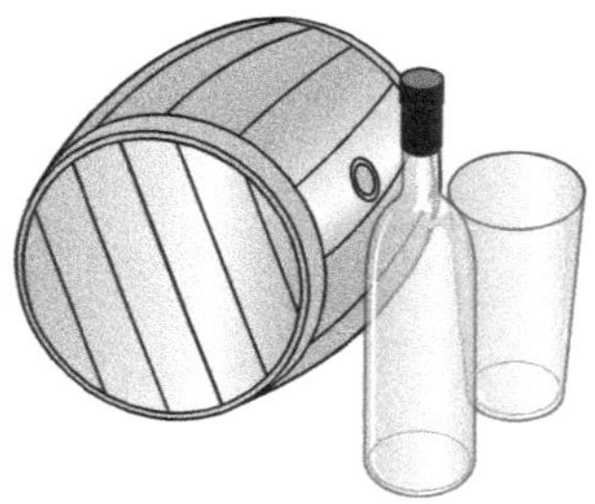

- minims
- pint
- gallon
- quarter
- barrel
- fluid drachm
- gill
- peck
- chaldron
- hogshead
- fluid ounce
- quart
- bushel
- firkin
- bottle (used to measure wine)

Around the World

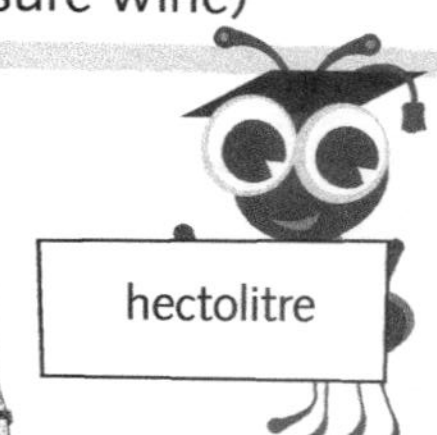

Most European countries measure volume and capacity using the metric system.

In the UK we use a combination of metric and imperial units.

The US uses imperial units. However, US and UK imperial measures are not all the same.

Investigate the differences between US and UK imperial units of measure for volume and capacity.

S&C Volume 5

The Maths Herald

Name: Date:

At Home

This table gives the average amount of water used for various household tasks.

Task	Litres
Bathing	90
Showering	27
Washing hands and face	3
Brushing teeth	1
Drinking	0·5
Flushing the toilet	9
Dishwasher	15 (per load)
Washing machine	118 (per load)

Approximately how much water does your family use in a day?

What about in a month?

What about in a year?

How does this compare with families of other children in your class?

What's the Problem?

You have these 3 buckets.

The 9-litre bucket is full of water.

The other 2 buckets are empty.

There are no markings on any of the buckets.

How could you measure every litre from 1 litre to 9 litres?

The Puzzler

Using 3 or more of the capacities above, and only addition and subtraction, make each of the following amounts.

475 LITRES 1837 LITRES 846 LITRES
2338 LITRES 1538 LITRES

Construct

Using 1 cm squared paper, cut out a 10 cm × 10 cm square.

Then cut out a 1 cm square from each of the 4 corners (the grey squares on the diagram on the right).

Now fold along the dotted lines and stick to make a box.

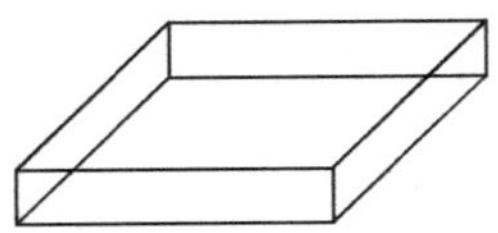

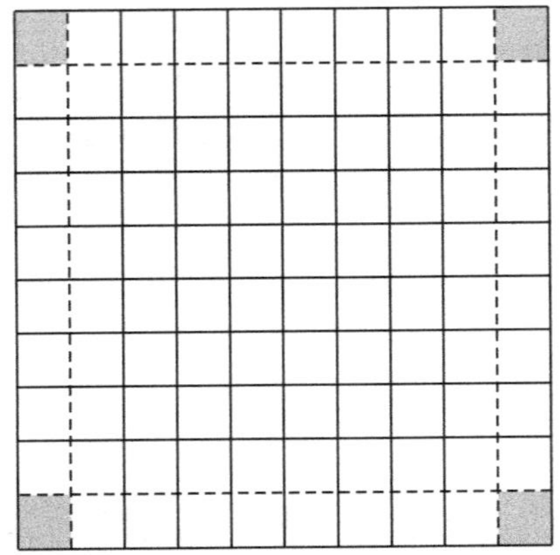

What is the volume of the box?

Repeat the above, this time cutting out a 2 cm square from each of the four corners.

What is the volume of the box?

Repeat the above twice more. For the third square cut out a 3 cm square from each of the 4 corners and for the fourth square cut out a 4 cm square from each of the 4 corners. Work out the volume of each box.

Which of your 4 boxes has the smallest volume?

Which of your 4 boxes has the greatest volume?

What if you tried with different sized squares?

At Home

Investigate the price of petrol and diesel from a range of different petrol stations.

If it takes 40 litres to fill up a car with petrol / diesel, how much would it cost at each of the different petrol stations?

What is the difference in price between the most and least expensive petrol / diesel?

Which is more expensive, petrol or diesel?

Issue 21 – Volume and capacity

In the Past

The Babylonians were the first people to introduce a standard system for measuring volume and capacity.

Investigate the following measures and how they are related:

- *sila*
- *gin*
- *she*

Let's Investigate

True or false?

1 litre of water weighs the same as 1 kilogram.

500 ml of water weighs the same as 500 g.

Conduct an experiment to find out.

Write about your experiment.

Provide evidence for your reasoning.

In the Past

The Ancient Greeks measured capacity differently depending on whether they needed to measure a dry or a wet (liquid) item.

The most common unit used was the *cotyle* or *cotyla*.

A larger unit was the *choenix*.

Both the Greeks and Romans used containers called *amphora*. They often used these for liquids. They had a special one called a *wine amphora*.

Find out about the Greek and Roman units for measuring volume and capacity.

Let's Investigate

A palindrome is a phrase, word or number that reads the same backwards as forwards. For example, 'MADAM', 'NURSES RUN', 575, 2002.

These 2 dates were both palindromic dates:

10 February 2001 (10022001)

20 February 2002 (20022002)

Using the date format DDMMYYYY what are some other palindromic dates in the 21st century?

At Home

In a year:

- how long do you spend watching television?
- how long do you spend on a computer or playing computer games?
- how long do you spend in bed?
- how long do you spend brushing your teeth?

Make up some 'in a year' statements of your own.

What's the Problem?

How long would it take you to count to 1 million?

Without actually counting to 1 million, how can you arrive at an approximate answer?

What's the Problem?

Lisa's great-grandma says that she is almost 30 000 days old and that she has slept for over 200 000 hours during her lifetime.

Could this be true? Why? / Why not?

What other 'amazing facts' might Lisa's great-grandma say?

S&C Volume 5

The Maths Herald

Name: Date:

Around the World

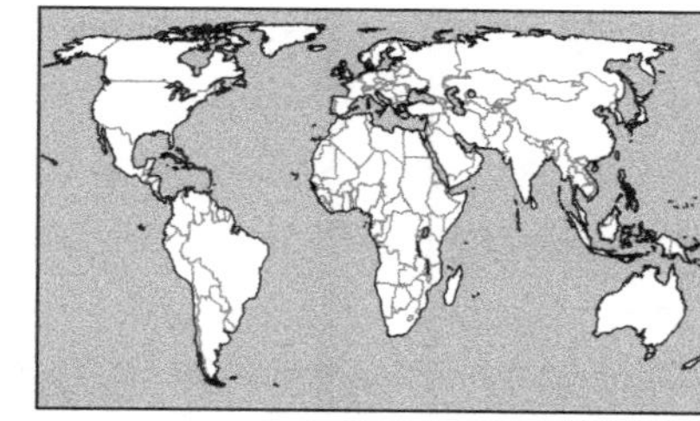

The world is divided into 24 time zones. The base time zone is at 0° longitude which passes through the Royal Greenwich Observatory in London. Each time zone spans 15° longitude although some borders are altered slightly to fit certain parts of the world. The time is set according to Greenwich Mean Time (GMT).

Find a map of the world that shows the different time zones.

Write about what children in other parts of the world might be doing as you are doing different things throughout the day. For example:

- when you get up in the morning
- when you start school
- when you're having lunch
- when you finish school
- when you're going to bed at night

Around the World

Draw a 24-hour timeline.

On your timeline mark and label the times that you do each of the following activities:

- Get up in the morning
- Start school
- Have lunch
- Finish school
- Go to bed at night

Imagine someone like you living in Sydney, Australia doing the same things as you throughout their day.

Using a different colour, mark and label on your timeline when they would be doing each of these things. Then see what you are doing when they are doing these activities.

Famous Mathematicians

Carl Gauss (1777–1855), was a German mathematician who devised a formula for finding out on which day of the week a person was born.

$$W = D + M + C + Y$$

W = Day of the week in which the person was born
D = Day of the month the person was born
M = Month in which the person was born
C = First 2 digits of the year in which the person was born
Y = Last 2 digits of the year in which the person was born
His formula also involved using the following tables.

M (month)			
Jan	Feb	March	April
0	3	3	6
May	June	July	Aug
1	4	6	2
Sept	Oct	Nov	Dec
5	0	3	5

C (first 2 digits of the year)		
15 …	19…	1
16…	20…	0
17…	21…	5
18…	22…	3

W (day of the week)						
Sun	1	8	15	22	29	…
Mon	2	9	16	23	30	…
Tue	3	10	17	24	31	…
Wed	4	11	18	25	32	…
Thu	5	12	19	26	33	…
Fri	6	13	20	27	34	…
Sat	7	14	21	28	35	…

Y (last 2 digits of the year)																			
00	06		17	23	28	34		45	51	56	62		73	79	84	90			0
01	07	12	18		29	35	40	46		57	63	68	74		85	91	96		1
02		13	19	24	30		41	47	52	58		69	75	80	86		97		2
03	08	14		25	31	36	42		53	59	64	70		81	87	92	98		3
	09	15	20	26		37	43	48	54		65	71	76	82		93	99		4
04	10		21	27	32	38		49	55	60	66		77	83	88	94			5
05	11	16	22		33	39	44	50		61	67	72	78		89	95			6

Example:
Beethoven was born on 16 December 1770.

$$\begin{aligned} W &= D + M + C + Y \\ &= 16 + 5 + 5 + 3 \\ &= 29 \end{aligned}$$

So Beethoven was born on a Sunday.

On which day of the week were you born?

Find out the year, month and date in which ten people were born. They can be friends, family members or famous people. However, they must have been born after 1582.

Use Gauss's formula to find out on which day of the week each person was born.

Let's Investigate

Here is another method for finding out on which day of the week a person was born.

Example: Beethoven was born on 16 December 1770.

Step 1: Write down the last 2 digits of the year in which the person was born. — 70

Step 2: Divide this number by 4, ignoring any remainders. — 17 (70 ÷ 4 = 17 r 2)

Step 3: Find the number for the month in which the person was born from the table below. — 6 (December)

Step 4: Write down the date in the month the person was born. — 16

Step 5: Take the century number of the year and divide it by 4.
If the remainder is 0 then write down 6
If the remainder is 1 then write down 4
If the remainder is 2 then write down 2
If the remainder is 3 then write down 0 — 4 (17 ÷ 4 = 4 r 1)

Step 6: Add together all the numbers from Steps 1–5. — 113 (70 + 17 + 6 + 16 + 4)

Step 7: Divide this answer by 7. — 113 ÷ 7 = 16 r 1

Step 8: Find the remainder in the table below to find out the day on which the person was born.

Months	
January	1
January (leap year)	0
February	4
February (leap year)	3
March	4
April	0
May	2
June	5
July	0
August	3
September	6
October	1
November	4
December	6

Remainder	Day
1	Sunday
2	Monday
3	Tuesday
4	Wednesday
5	Thursday
6	Friday
0	Saturday

Remainder 1 means that Beethoven was born on a Sunday.

Use this method to check your day of birth and the ten other people you chose in the Famous Mathematicians activity on page 2.

Focus on Science

Look through a newspaper that shows both the temperature and the wind speed. Choose different locations and, using the Wind-chill Factor Chart in the Focus on Science activity on page 3, work out what the temperatures actually feel like.

Record your results in a table.

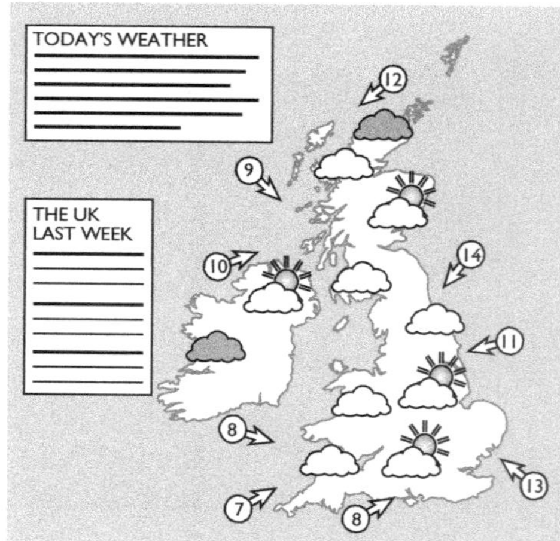

Focus on Science

Temperature is generally measured in either degrees Celsius (°C) or in degrees Fahrenheit (°F). Today, in the UK most temperatures are measured in degrees Celsius. This activity looks at temperatures in degrees Fahrenheit.

Over the course of a normal day your blood temperature can range between about 36·3°C (97·35°F) and about 37·5°C (99·52°F).

The table below shows, in degrees Fahrenheit, measurements of normal blood temperature over a 24-hour period.

Time	00:00	01:00	02:00	03:00	04:00	05:00	06:00	07:00	08:00	09:00	10:00	11:00
°F	98·4	97·9	97·5	97·45	97·35	97·35	97·5	97·7	97·9	98·2	98·7	99·3

Time	12:00	13:00	14:00	15:00	16:00	17:00	18:00	19:00	20:00	21:00	22:00	23:00
°F	99·4	99·3	99·3	99·45	99·5	99·55	99·5	99·4	99·3	99·1	98·9	98·65

Draw a line graph that shows the change in blood temperature over 24 hours.

Use squared paper to draw your graph.

You will need to think carefully about how to label the temperature axis.

S&C Volume 5

The Maths Herald

Name: Date:

Focus on Science

The *Gulf Stream* is a powerful warm current that starts in the Gulf of Mexico and crosses the Atlantic Ocean towards Europe. Due to this current, winters in Europe are much milder, with higher temperatures than they would be otherwise. If the current were to stop, European winter temperatures would be as low as North American temperatures at similar latitudes.

Investigate how different winter temperatures would be where you live if the Gulf Stream were to stop.

What about other major cities in Europe?

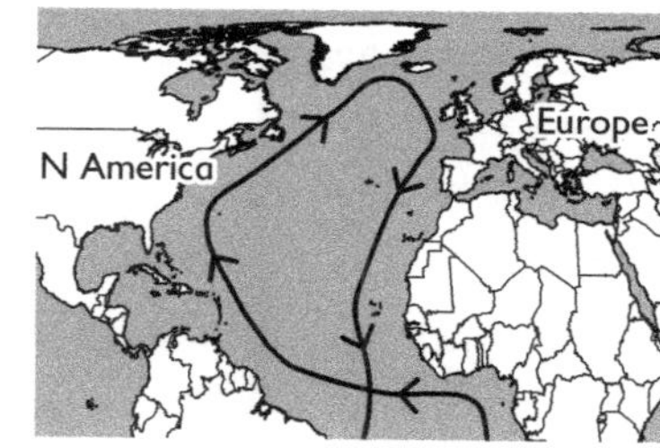

Clue:

Lines of latitude are imaginary lines joining points on the Earth's surface that are all of equal distance north or south of the equator.

Focus on Science

Choose 3 different locations inside and outside the school that you predict will give different temperature readings throughout the day.

Every hour throughout the day measure and record the temperature at each of these 3 locations.

Display your results in a table.

Were your predictions correct?

Repeat the experiment on other days.

Write a report comparing your results over different days.

Around the World

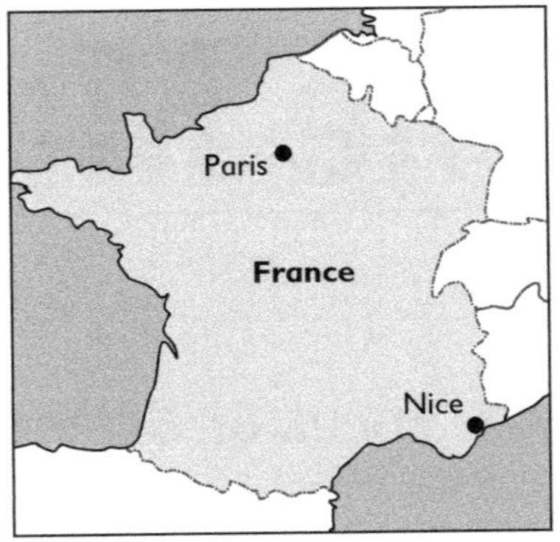

The table below shows the average daily maximum temperatures for 2 cities in France – Paris, in the north; and Nice, in the south.

Use the information in the table to write about the following:

- the temperature differences at different times throughout the year in Paris
- the temperature differences at different times throughout the year in Nice
- the temperature differences between Paris and Nice at different times throughout the year.

	Jan	Feb	Mar	Apr	May	June	July	Aug	Sept	Oct	Nov	Dec
Paris	8	7	10	16	17	23	25	26	21	17	12	8
Nice	12	12	14	19	21	27	28	28	25	22	17	14

Temperatures in degrees Celsius (°C)

Around the World

Use the information in the table in the Around the World activity above to draw a line graph that shows the average daily maximum temperatures for Paris and Nice - 1 line for each.

Use squared paper to draw your graph.

Technology Today

Use the information in the table in the Around the World activity above to draw another line graph for Paris and Nice. However this time use ICT to create the graph.

At Home

When cooking or baking something in an oven, recipes usually give the recommended temperature at which the food should be cooked.

Most often the temperatures are given in degrees Celsius or degrees Fahrenheit. However, sometimes a recipe will use terms such as 'bake in a moderate oven' or 'cook at a high temperature'.

Find out about different cooking temperatures and what names are given to each of these temperature ranges.

Record your results in a table. Include temperatures both in degrees Celsius and degrees Fahrenheit.

Focus on Science

Wind-chill or *wind-chill factor* causes people to feel colder because the wind affects the rate of heat loss from the body and has the effect of making it feel colder.

For example, when the temperature is 8°C and you're running along at 10 km/h into a headwind that is blowing at 30 km/h (total 40 km/h), it will feel as though the temperature is 3·3°C.

Wind-Chill Factor Chart

Temperature °C

Wind speed (km/h)	10	9	8	7	6	5	4	3	2	1	0	-1	-2	-3	-4
5	9·8	8·6	7·5	6·4	5·2	4·1	2·9	1·8	0·7	-0·5	-1·6	-2·7	-3·9	-5·0	-6·1
10	8·6	7·4	6·2	5·0	3·9	2·7	1·5	0·3	-0·9	-2·1	-3·3	-4·5	-5·7	-6·9	-8·1
15	7·9	6·7	5·4	4·2	3·0	1·7	0·5	-0·7	-2·0	-3·2	-4·4	-5·6	-6·9	-8·1	-9·3
20	7·4	6·1	4·9	3·6	2·3	1·1	-0·2	-1·5	-2·7	-4·0	-5·2	-6·5	-7·8	-9·0	-10·3
25	6·9	5·7	4·4	3·1	1·8	0·5	-0·8	-2·1	-3·3	-4·6	-5·9	-7·2	-8·5	-9·8	-11·1
30	6·6	5·3	4·0	2·7	1·4	0·1	-1·3	-2·6	-3·9	-5·2	-6·5	-7·8	-9·1	-10·4	-11·7
35	6·3	4·9	3·6	2·3	1·0	-0·4	-1·7	-3·0	-4·3	-5·6	-7·0	-8·3	-9·6	-10·9	-12·2
40	6·0	4·6	3·3	2·0	0·6	-0·7	-2·0	-3·4	-4·7	-6·1	-7·4	-8·7	-10·1	-11·4	-12·7
45	5·7	4·4	3·0	1·7	0·3	-1·0	-2·4	-3·7	-5·1	-6·4	-7·8	-9·1	-10·5	-11·8	-13·2
50	5·5	4·1	2·8	1·4	0·0	-1·3	-2·7	-4·1	-5·4	-6·8	-8·1	-9·5	-10·9	-12·2	-13·6

Work with a friend and take turns to say what the temperature feels like. The other person has to say what the temperature is and how strong the wind is. For example, if the temperature feels like -5·2°C, then the temperature could be 0°C with a wind of 20 km/h; or 1°C with a wind of 30 km/h.

What's the Problem?

Look at this diagram.

ABCD are the 4 corners of a square.

Square EFGH was made by joining together the 4 midpoints of square ABCD.

Square IJKL was made by joining together the 4 midpoints of square EFGH.

Square MNOP was made by joining together the 4 midpoints of square IJKL.

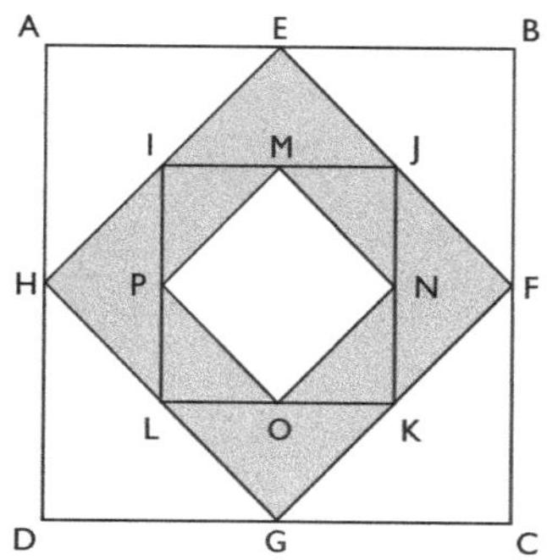

If the area of square ABCD is 36 cm², what is the area of the shaded space?

Let's Investigate

Investigate how much paint it would take to paint all the walls in your classroom.

Find out the cost of such a project.

How does this compare with the cost of covering the walls with wallpaper?

Which would be the better option? Give reasons for your conclusions.

The Puzzler

ABF is an isosceles triangle. BF = AG = 6 cm.

If FB is half the length of BC and ABF = CDE, then calculate each of the following:

- The area of rectangle BCEF
- The area of triangle CEF
- The total area of the shape.

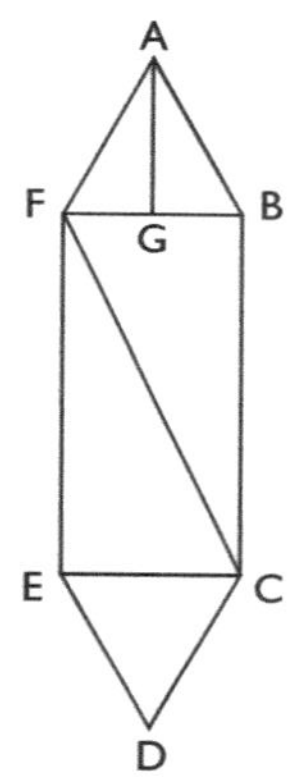

What's the Problem?

Look at the diagram in the What's the Problem? activity above.

What is the area of triangle AEH?

S&C Volume 5

The Maths Herald

Name: Date:

The Puzzler

The diagonals of a kite are 3 cm and 8 cm long.

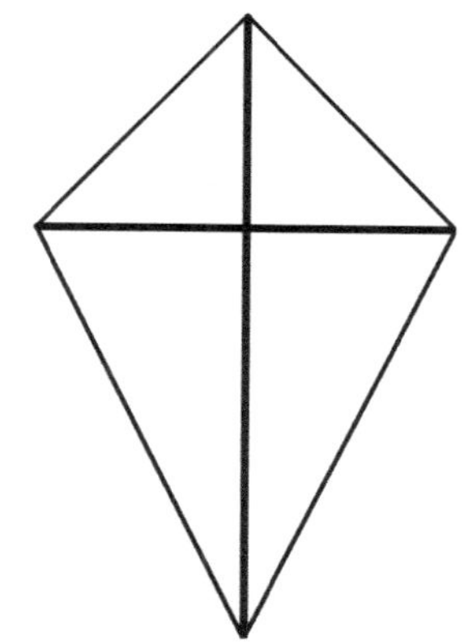

What is the area of the kite?

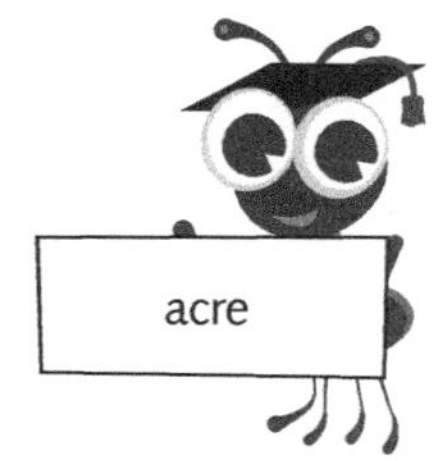

How did you work out the answer?

The Puzzler

Arrange 12 matchsticks like this:

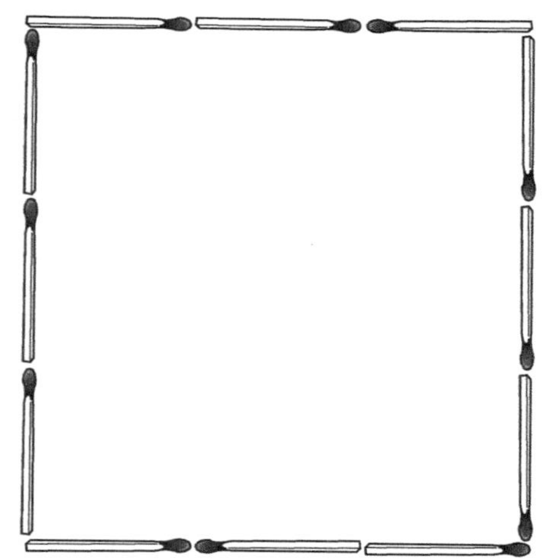

Then place 7 matchsticks inside the square to divide it into 3 shapes all with the same area.

Draw your result.

These 12 matchsticks enclose an area of 9 square units.

Rearrange the matchsticks to make shapes with the following square units:

8, 7, 6, 5, 4, 3.

Draw your result.

Around the World

The *Peters Projection* world map is named after Dr Arno Peters. Unlike a conventional map of the world, it shows the relative size of different countries to scale.

Using a Peters Projection map, choose 2 countries which, by sight, are a similar size. Trace the outline of these 2 countries on to graph paper and count the number of squares. Which is the larger country by area?

Choose other countries and compare their areas.

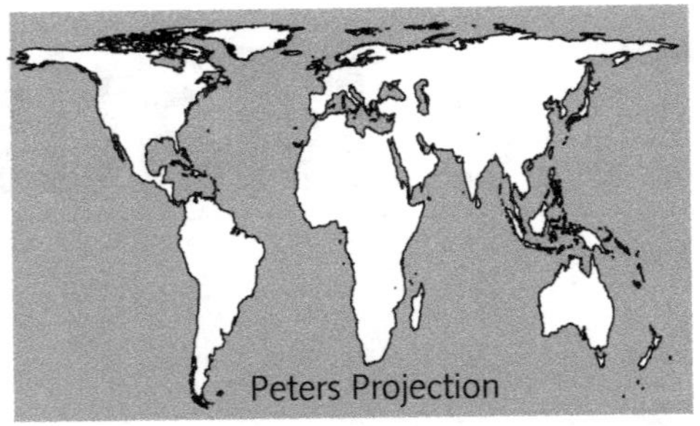

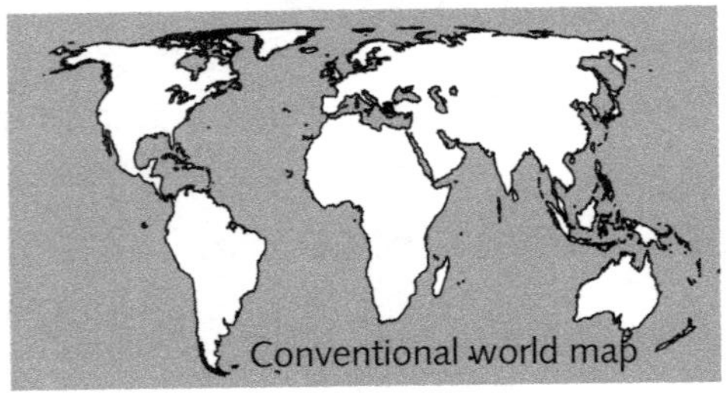

Focus on Science

How would you find the area of your foot?

What about your hand?

Design a reliable test to measure the area of feet and hands.

Write about your test and how reliable it is as a measure of the area of feet and hands.

What is the average foot and hand area of your class?

Who is above and below the class average?

Is there any correlation between foot and hand area?

Construct

Using 1 cm squared paper, draw a series of different sized quadrilaterals.

Do not follow the lines on the squared paper, but do make sure that your lines are straight and the vertices are on points where vertical and horizontal lines of the paper cross.

What is the area of each quadrilateral?

Describe how you calculated the areas.

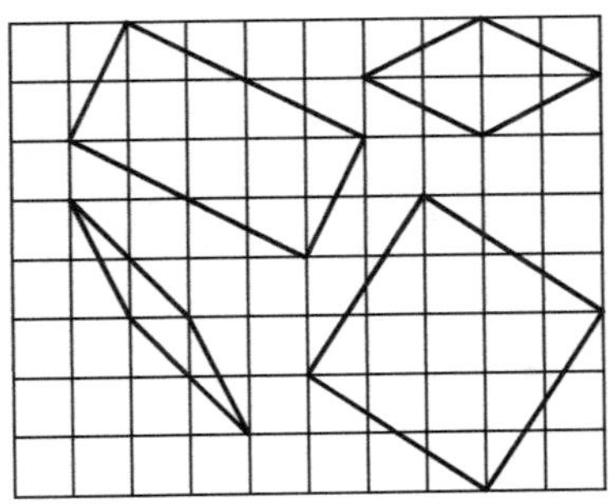

What's the Problem?

These 3 cuboids all show the areas of the faces, but not the cuboids' dimensions.

What is the length, height and width of each cuboid?

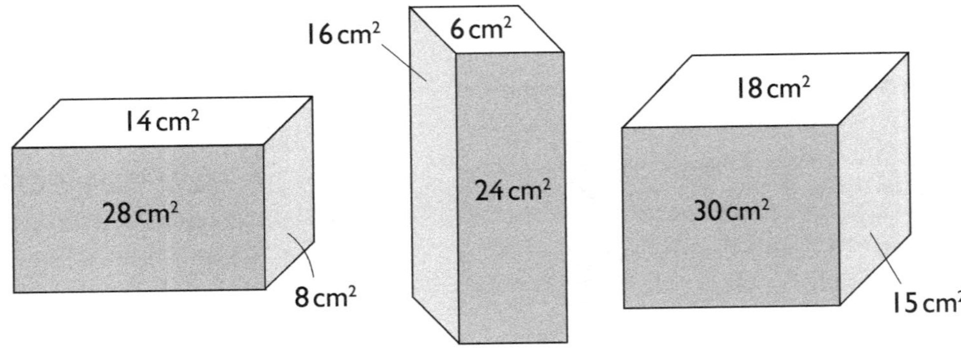

What's the Problem?

Richard is re-tiling his bathroom.

This is the calculation he has done to find the surface area of the 4 walls of his bathroom:

$2(3\text{ m} \times 2{\cdot}5\text{ m}) + 2(2\text{ m} \times 2{\cdot}5\text{ m}) =$

Complete Richard's calculation.

He is using 20 cm square tiles.

How many tiles does Richard need to re-tile his bathroom?

Tiles are sold in boxes of 25. How many boxes will Richard need to buy?

A box of tiles costs £12.50. How much will Richard have to pay?

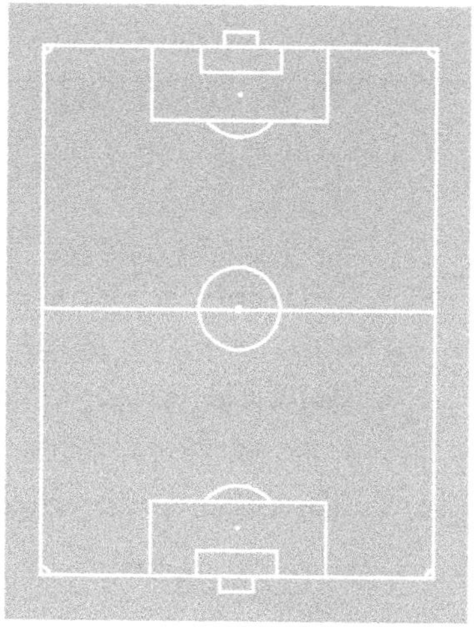

Sports Update

Find out all the dimensions of a professional football pitch.

If you were asked to paint the lines that form the perimeter of the pitch, how many metres would you have to paint?

How many metres would you have painted if you painted all the lines on a football pitch?

Ignore any circles or parts of circles.

Let's Investigate

It is easy to find the perimeter of rectilinear shapes like the ones shown below:

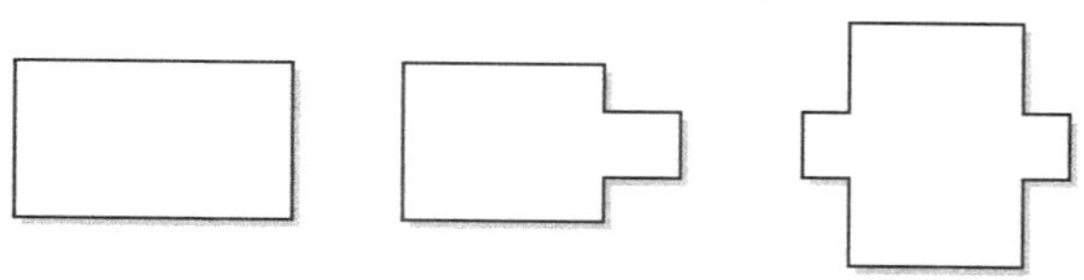

How would you go about finding the perimeters of non-rectilinear shapes? Write about your conclusions and provide examples to justify your decisions.

The Puzzler

Look at this rectangle. It is made up of 12 small squares.

- Can you remove 1 square so that the area is reduced but the perimeter stays the same?
- What is the maximum number of squares you can remove without changing the perimeter?

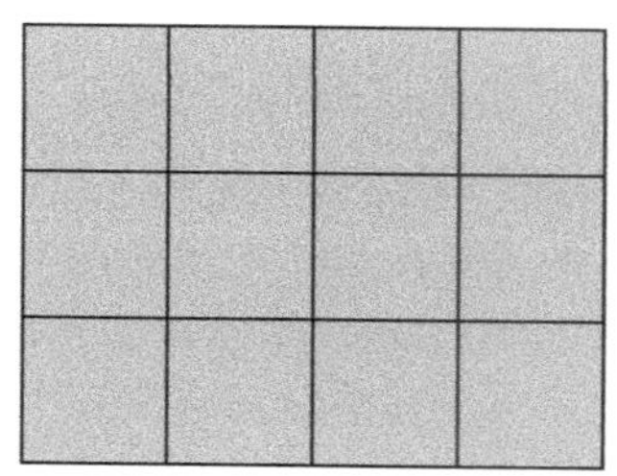

S&C Volume 5

The Maths Herald

Name: Date:

What's the Problem?

A swimming pool is 12 m long and 5 m wide.

Around the pool is a 0·5 m border.

What is the perimeter of the outside of the border?

What is the area of the border?

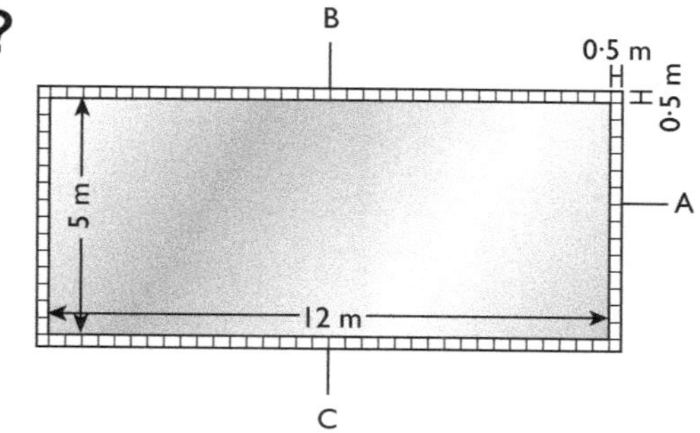

What's the Problem?

Rectangle	A	B	C	D	E
Area	48 cm^2	70 cm^2	90 cm^2	96 cm^2	144 cm^2
Perimeter	32 cm	34 cm	42 cm	56 cm	48 cm

Paul found the area and perimeter of 5 rectangles.

Find the dimensions of each rectangle.

Let's Investigate

- Roll a 0–9 dice twice.
- Multiply together the 2 numbers rolled.
- Using squared paper, draw a shape with this size perimeter.
- Can you draw other shapes with the same perimeter?
- Repeat, rolling the dice again.

These 4 shapes each have a perimeter of 18 units.

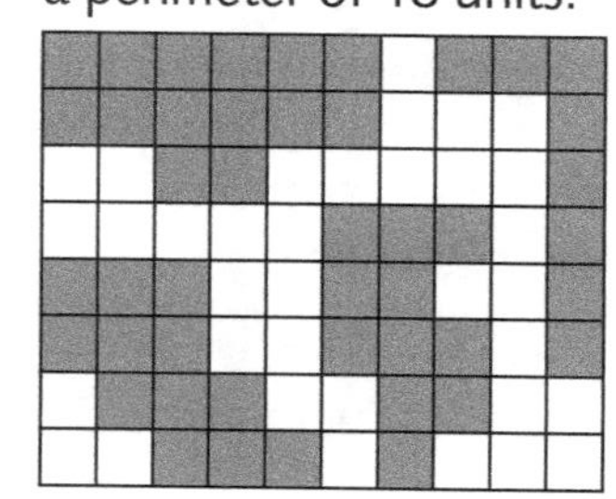

What's the Problem?

The following diagram shows the area and perimeter of 10 allotments.
Use this information to calculate the dimensions of each allotment.

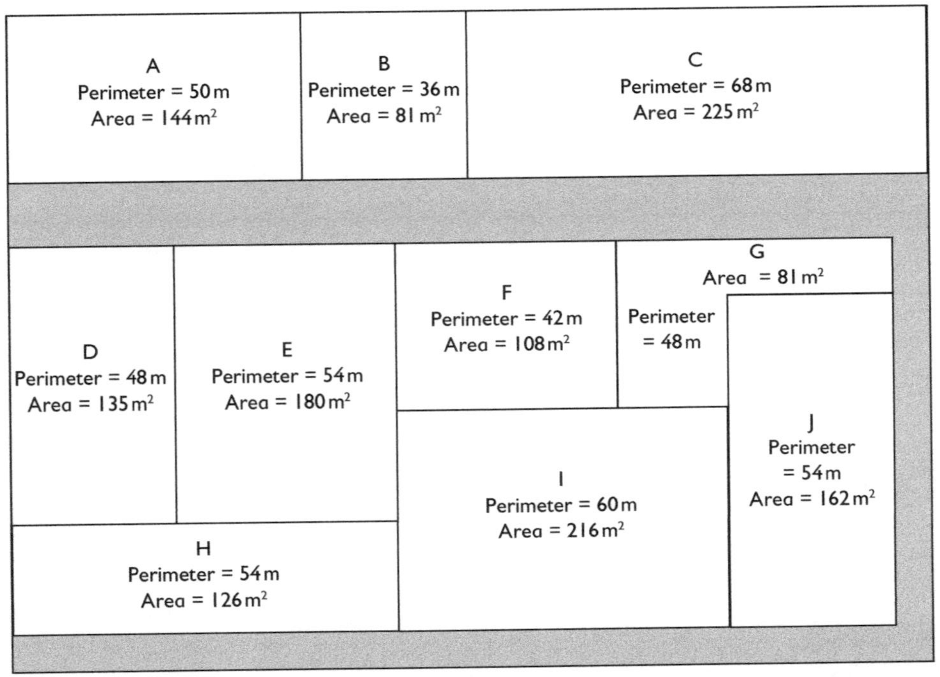

What's the Problem?

Given the total size of the 10 allotments in the What's the Problem? activity above, what other sizes could the 10 allotments be?

Draw your diagram on squared paper.

Be sure to include all the dimensions on your diagram.

Let's Investigate

Use a trundle wheel to find the perimeter of your school playground.

How long does it take you to walk around it?

If you were trying to keep fit by running for half an hour every day, how many circuits of the perimeter would you have to do?

How far would you run?

Let's Investigate

- Investigate drawing shapes that have areas of $24\,cm^2$ but different perimeters.
- Investigate drawing shapes that have perimeters of 24 cm but different areas.

Let's Investigate

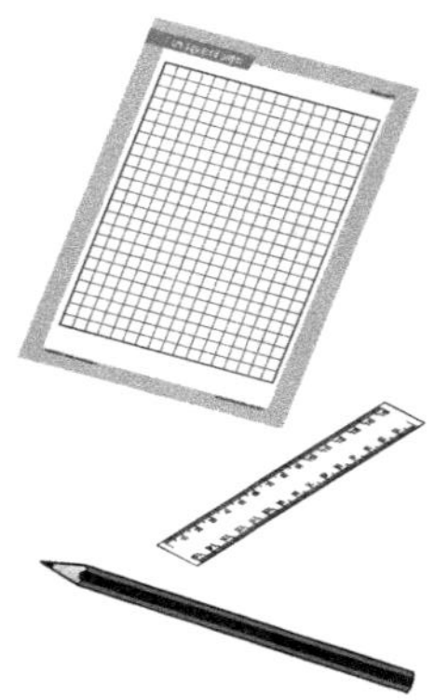

Using mathematical paper, draw a small polygon.

Label it 'A' and work out its area and perimeter.

Draw shape 'A' again but with each side twice as big.

Label it 'B' and work out its area and perimeter.

Now draw shape 'A' again but this time with each side three times as long.

Label it 'C' and work out its area and perimeter.

What do you notice about the areas and perimeters of shapes A, B and C?

Construct

Herne Developers have just built a new block of flats. *Meroo Meadows* offers the following accommodation:

- 5 one-bedroom flats, each of 45 m^2
- 7 two-bedroom flats, each of 60 m^2
- 3 three-bedroom flats, each of 90 m^2
- 1 penthouse of 150 m^2

Meroo Meadows

Luxury 1, 2 and 3 bedroom flats and penthouse boasting spectacular lake views.

Design 4 floor plans: 1 for each sized flat.
Your plans must include the following features:

1-bedroom flats include:
- living area
- kitchen
- bathroom

2-bedroom flats include:
- living area
- kitchen
- bathroom
- en-suite bathroom for master bedroom

3-bedroom flats include:
- living area
- kitchen
- bathroom
- en-suite bathroom for master bedroom

The penthouse includes:
- 4 bedrooms each with en-suite bathroom
- 2 living areas
- kitchen
- dining area
- bathroom

Money Matters

Investigate what prices Herne Developers could ask for each of the flats and penthouse in *Meroo Meadows*.

Be prepared to justify your pricing structure.

Meroo Meadows

Prices

1 bedroom flat £
2 bedroom flat £
3 bedroom flat £
Penthouse £

S&C Volume 5

The Maths Herald

Name: Date:

Sports Update

In order for the body to work properly, it requires an output of energy. This energy comes from the body's resources and through the food we eat. The energy used is measured in kilojoules per minute per activity (kJ/min).

This table lists various activities and the average energy used for each.

Make a list of all the activities you do in a 24-hour period.

Work out how many minutes of each activity you did.

Use the table to calculate how many kilojoules of energy you used in the 24-hour period.

Activity	kJ/min
Sleeping	4·2
Watching TV / On the computer	5
Eating	5·9
Dressing	9·7
Making the bed	16·4
Having a bath/shower	17·6
Walking	22·7
Playing football or similar sport	36·5
Running	52·9

Let's Investigate

A palindrome is a word, phrase or number that reads the same backwards as forwards.

For example, 'MADAM', 'NURSES RUN', 575, 2002.

These 2 times are both palindromic times on a 12-hour digital clock:

3:33 2:42

How many palindromic times are there on a 12-hour digital clock?

Do you think there are more or less palindromic times on a 12-hour digital clock than on a 24-hour digital clock?

How many palindromic times are there on a 24-hour digital clock?

Around the World

On what date and at what time does Allison's plane depart?

Around the World

At what altitude does an aeroplane fly?

Is this different for planes of different sizes?

Write statements comparing the altitude of an aeroplane to the height of a tall building or structure, for example the Eiffel Tower or the London Eye.

Around the World

How far is it from London to Sydney?

How long does it take a plane to fly this distance?

On average, how far does the plane travel each hour?

If you could run that fast, approximately how long would it take you to travel to school?

Show how you calculated your answers.

Focus on Science

Work with a partner.

How good are you at estimating time?

Devise an experiment to find out. Think carefully about what things you will need to ensure a high degree of accuracy.

Take turns to estimate how long 10 seconds is.

Repeat several times each, recording your results.

Write about the differences in your estimates.

Now try for longer periods, such as: 30 seconds, 1 minute and 2 minutes.

Do the differences in your estimates increase, decrease or stay the same as the amount of time increases?

Write about the reasons for any differences in your estimates.

Do you think that adults' estimates are similar to yours? Find out.

Try a range of different aged adults.

Construct

You have been invited to a birthday party and are taking a present in a box.

How could you work out how much wrapping paper you would need?

Using newspaper, work out an efficient method for any cuboid-shaped present.

Write about your method.

Let's Investigate

Gather together a collection of cylinders and different types of small objects such as counters, paper clips, Centicubes and marbles.

Fill each of the containers with the small objects. Use only 1 type of object in each container.

Record how many or how much it takes to fill each container.

Write a description of how to find the volume of a cylinder.

Construct

By joining together some or all of these shapes, what other polygons can you make?

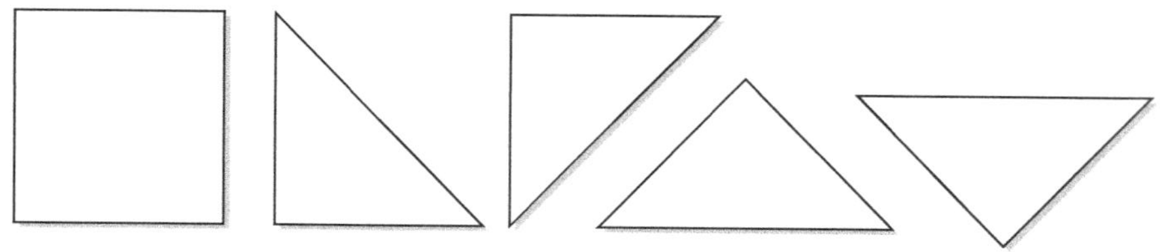

Draw your results.

Let's Investigate

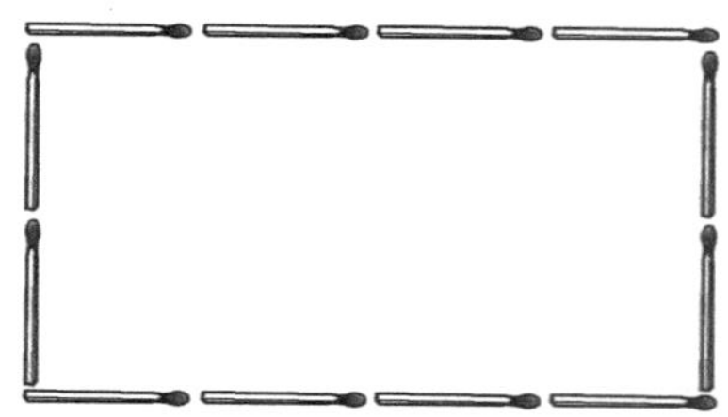

Make a polygon using 12 matchsticks.

All the angles of your polygon must be right angles.

Make as many different polygons as you can.

What is the perimeter of each shape?

What about polygons with angles of different sizes?

Let's Investigate

Look at this pattern. It is made from 7 identical circles.

A *point of intersection* is a place where 2 or more lines cross each other.

Find as many different triangles as you can by joining 3 points of intersection.

Find as many different quadrilaterals as you can by joining 4 points of intersection.

Record your results on Resource sheet 8.

S&C Volume 5

The Maths Herald

Name: Date:

Famous Mathematicians

Pythagoras was a Greek philosopher (569–500 BC). He made many important discoveries in mathematics, particularly in geometry. One of Pythagoras's greatest discoveries is called the *Pythagorean theorem*.

The longest side of a right-angled triangle, the side opposite the right angle, is called the *hypotenuse* (labelled c).

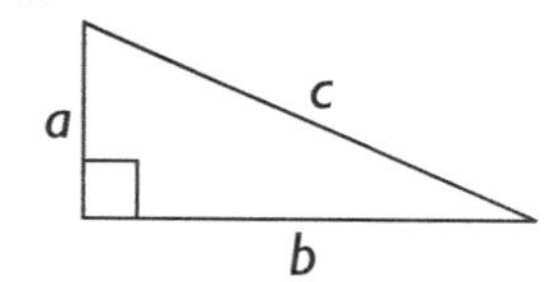

Pythagoras's theorem states that, for any right-angled triangle, the square of the length of the hypotenuse is equal to the sum of the squares of the lengths of the other 2 sides.

$$c^2 = a^2 + b^2$$

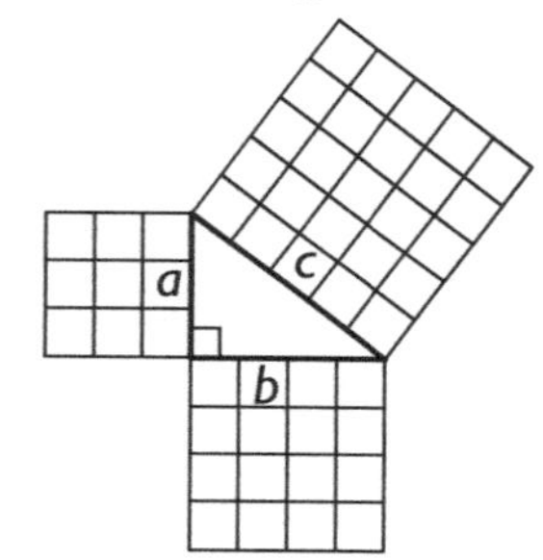

Explain how this diagram shows Pythagoras's theorem.

The Puzzler

Draw a square of any size.

Mark points A and B – they are in the middle of the sides of the square.

Draw the 2 lines and cut the square into 4 pieces.

Fitting 2 or 3 of these pieces at a time together, draw and name as many shapes as you can.

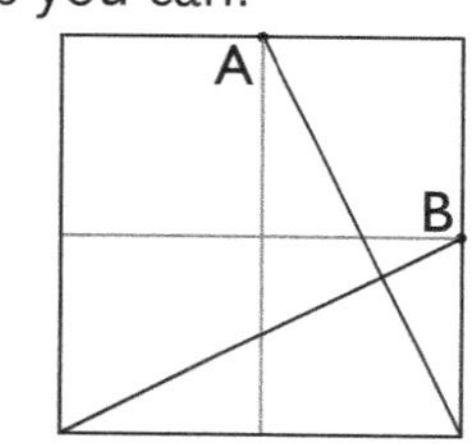

Famous Mathematicians

Cut out the puzzle from Resource sheet 7 and arrange the 5 grey pieces to make one square on the hypotenuse.

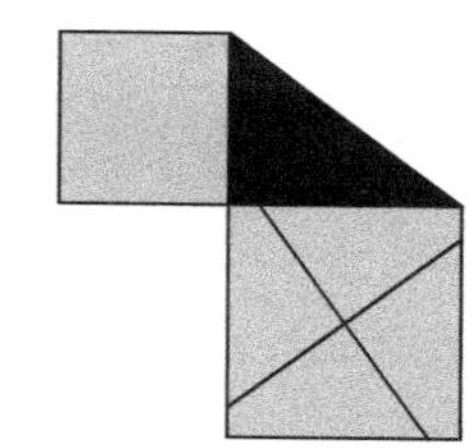

Explain how this shows Pythagoras's theorem.

The Language of Maths

What is a quadrilateral?

How many different types of quadrilaterals are there?

Design a poster about quadrilaterals. On your poster include a drawing and a brief definition of each type of quadrilateral.

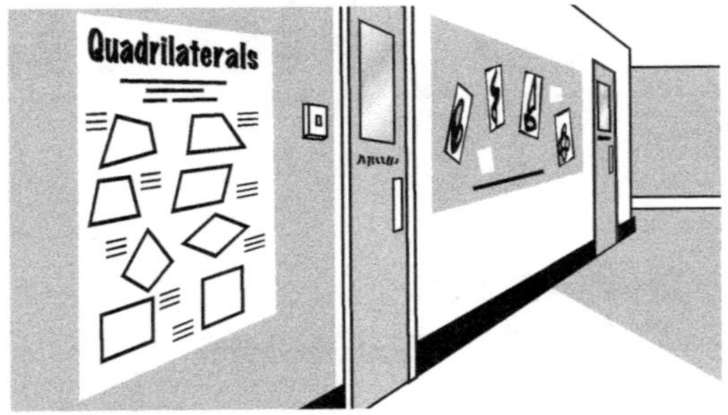

What's the Problem?

Copy an enlarged version of this grid.

Then draw at least 1 quadrilateral in each box and name each one.

Number of pairs of equal sides \ Number of right angles	0	1	2 or more
0			
1			
2			

Let's Investigate

By overlapping 2 congruent squares an irregular pentagon has been formed.

Can you make any of these shapes in the same way?

- rectangle
- rhombus
- kite
- isosceles triangle
- decagon
- hexagon
- octagon
- trapezium

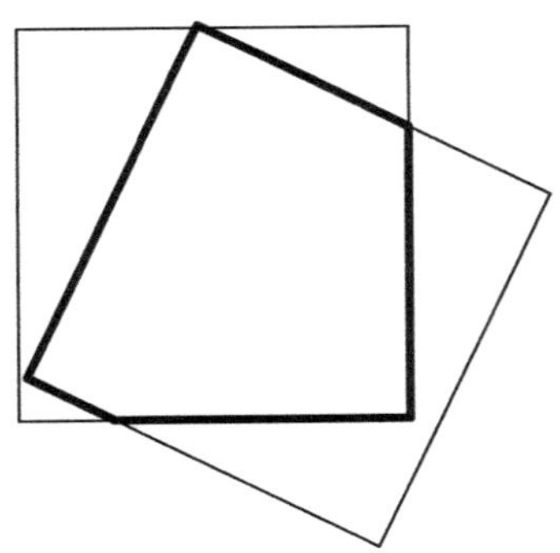

What shape can't you make? Why?

What shapes can you make by overlapping 3 squares?

Let's Investigate

How many different quadrilaterals can be drawn on a 3 × 3 dot grid?

Here is one.

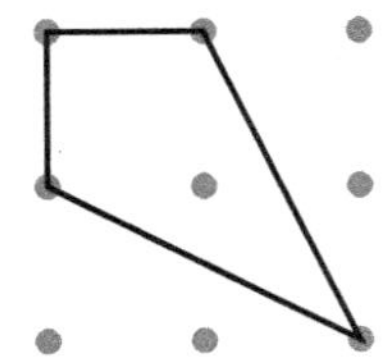

Draw your quadrilaterals on squared dot paper.

Be sure not to include different orientations of the same quadrilateral, for example:

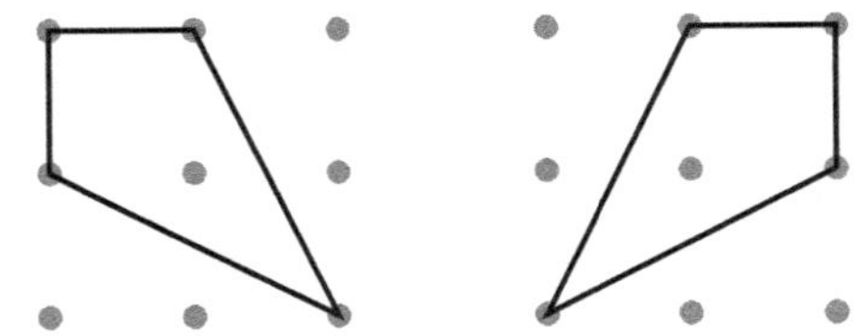

When you have made as many different quadrilaterals as possible, go back and name each one.

Let's Investigate

Use a ruler to draw any quadrilateral.

Measure the length of each side of the quadrilateral and mark the midpoints.

Then use a ruler to draw 4 lines that connect the midpoints of adjacent sides.

Repeat for a range of different quadrilaterals.

Write about what you notice.

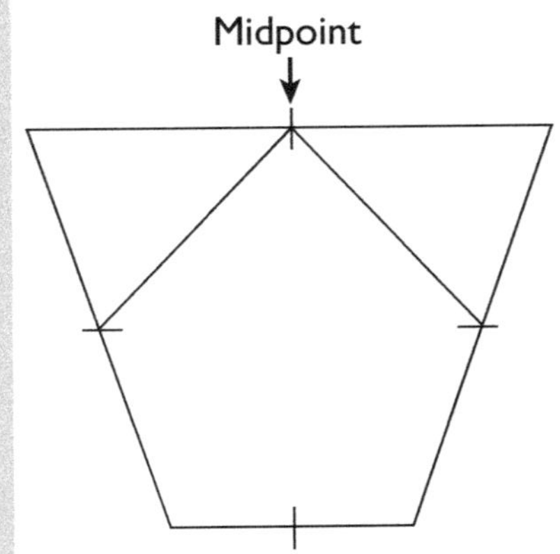

Let's Investigate

Using squared dot paper, draw a box around a 3 × 3 square.

Investigate the polygon you can make with the greatest number of sides.

What about for a 4 × 4 or 5 × 5 square?

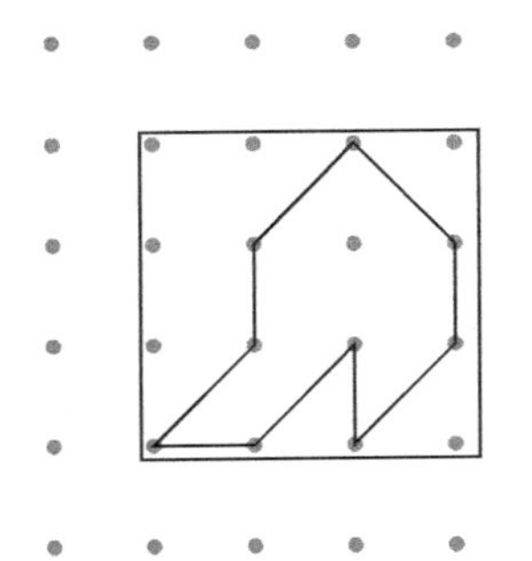

Looking for Patterns

Face area refers to the area of 1 face on a cube.

It is calculated by multiplying the length of the cube's face by its width to give square units.

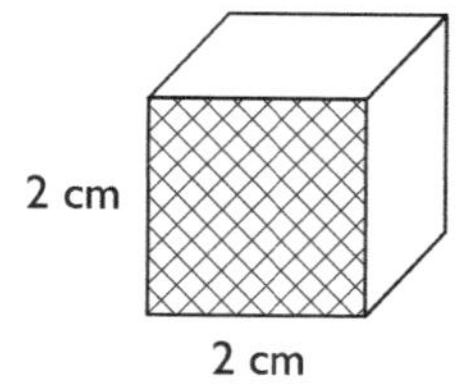

Face area

= 2 cm × 2 cm

= 4 cm^2

Surface area refers to the total area of all 6 faces on a cube.

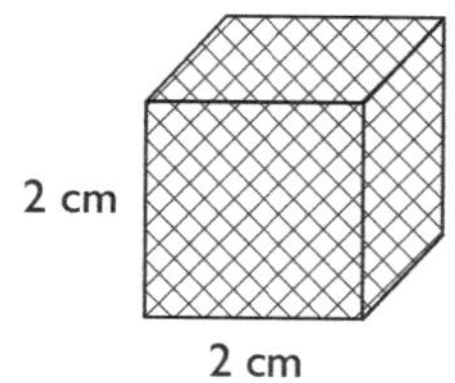

Surface area

= (2 cm × 2 cm) × 6

= 24 cm^2

Volume refers to the amount of space taken up by a cube.

It is calculated by using the formula length × width × height (L × W × H). The answer is expressed in cubic units.

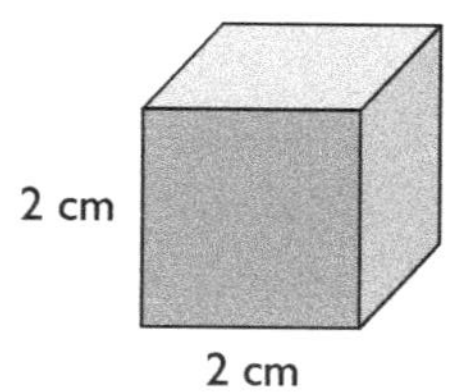

Volume

= L × W × H

= 2 cm × 2 cm × 2 cm

= 8 cm^3

Work out the face area, surface area and volume for cubes from 1 cm to 10 cm.

Record your results in a table.

What patterns do you notice?

Construct

Carefully draw a regular pentagon with sides of 4 cm.

Cut out the pentagon and use this as a template for drawing the net of a regular dodecahedron as shown here.

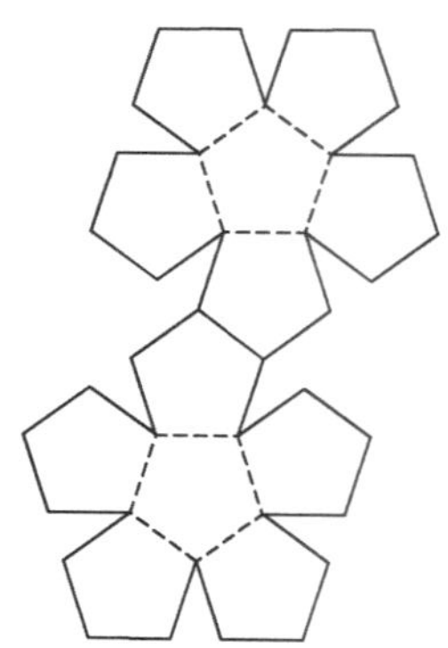

Glue or tape the edges together to make a solid.

Can you use this method to make other solid shapes such as an icosahedron or an octahedron?

S&C Volume 5

The Maths Herald

Name: Date:

Let's Investigate

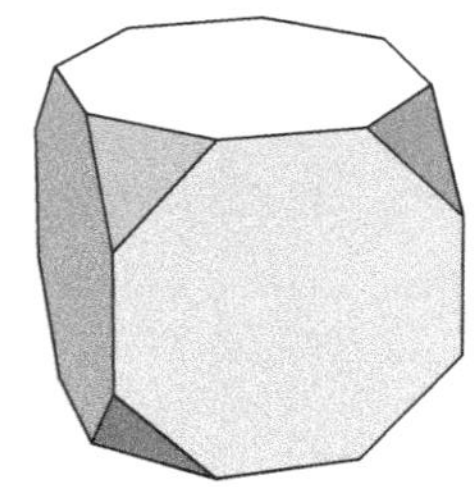

A *truncated solid* is one where the corners have been cut off.

This is an example of a truncated cube.

What are the different faces of this truncated cube?

How many faces, vertices and edges does a truncated cube have?

Investigate how many faces, vertices and edges other truncated solids have.

Let's Investigate

All the faces of this cube have the number 1 written on them.

How many different arrangements can you make if each face either has the number 1 or 2 written on it? Ignore the orientation of the numbers.

What about the numbers 1, 2 or 3?

The Puzzler

The diagrams are of 3-D shapes when looked at from directly above.

What could each of these 3-D shapes be? Write a description to justify your reasoning.

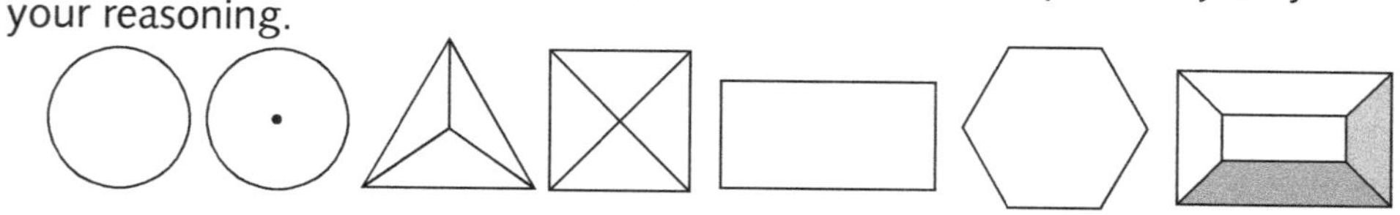

Construct

Plan and draw a net to make this cuboid.

Then construct the cuboid.

Look at your completed cuboid.

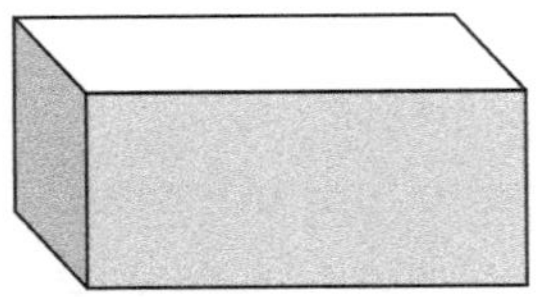

Now plan and draw a net to make this shape.

Then construct the shape.

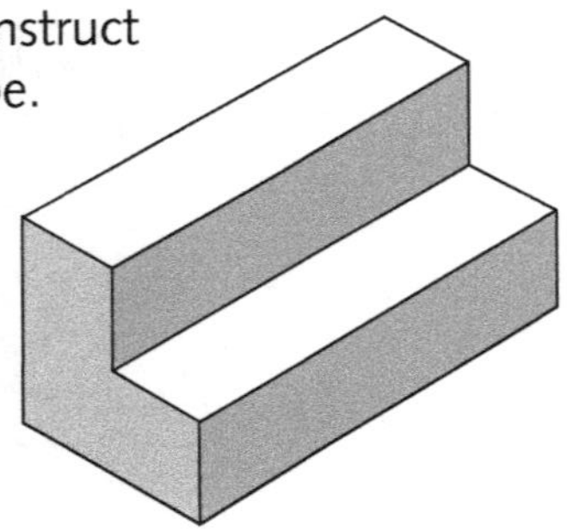

Construct

Plan and draw a net to make each of these 3-D shapes.

Then construct each of the 3-D shapes.

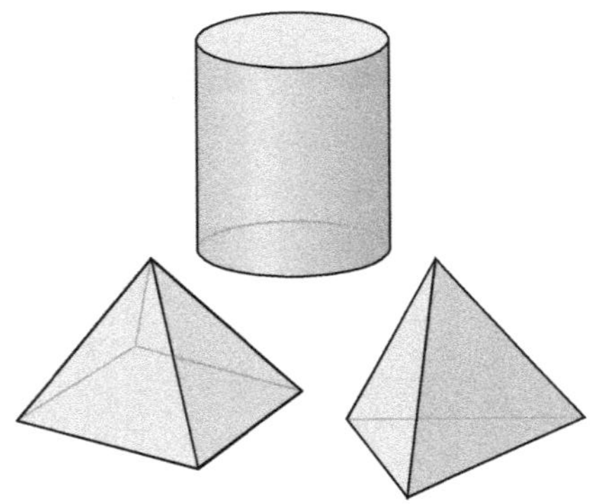

When you have made a 3-D shape, write about how successful your net was.

Are there different nets you could make that would create the same 3-D shape?

The Language of Maths

Polyhedrons are 3-D shapes with many faces.

They were first described over 2000 years ago by a Greek mathematician called Euclid.

There are 5 regular or perfect polyhedrons (also known as *Platonic solids*).

The faces of a perfect polyhedron are the same shape and size. That is, they are *congruent*.

What are the names of each of the 5 perfect polyhedrons?

Find images for each one.

The Language of Maths

What is a prism?

There are 4 common prisms. They are:

- triangular prism
- quadrangular prism
- hexagonal prism
- octagonal prism.

Find images for each one.

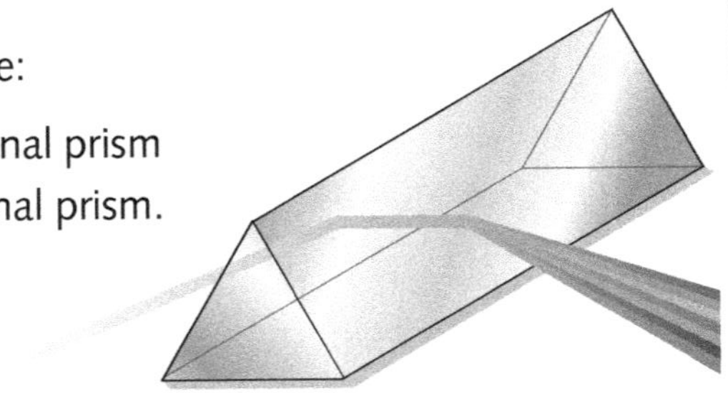

Construct

Isometric paper can be used to draw 2-D representations of 3-D shapes.

Look at the drawing of this cube.

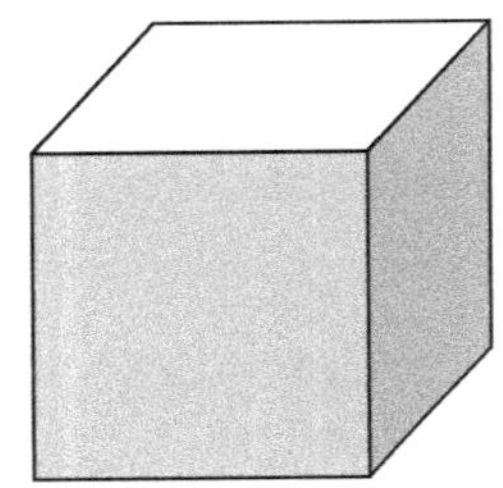

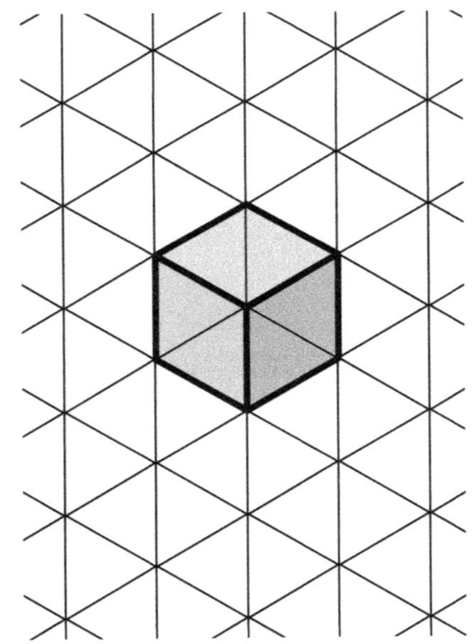

Using isometric paper, draw the cube.

Then draw these shapes on isometric paper.

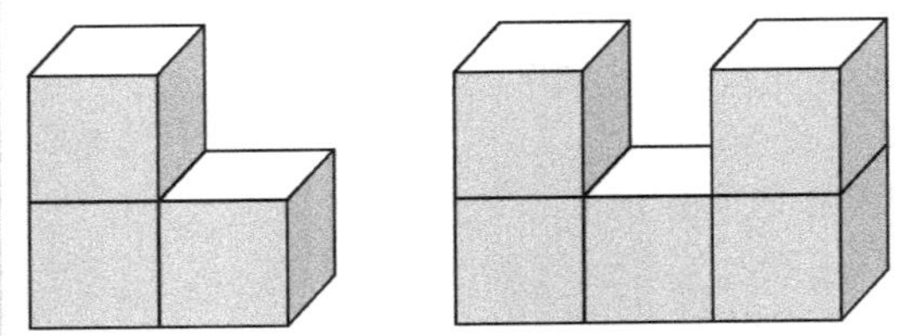

Use Centicubes to make your own models and then draw them on isometric paper.

Let's Investigate

Write down all the letters of the alphabet in capitals.

Investigate the different types of angles that are in each of your letters. For example,

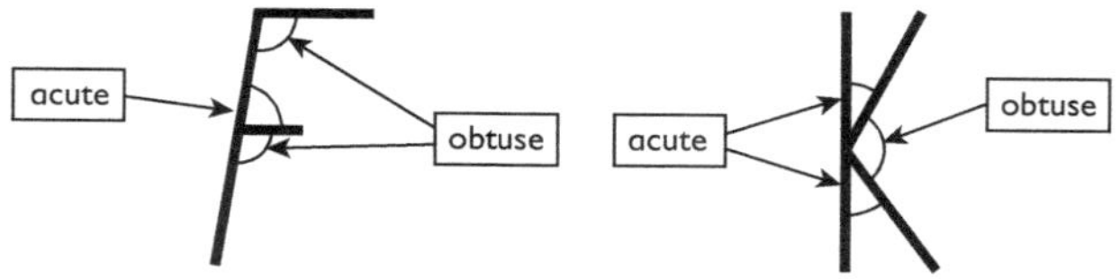

Now investigate capital letters created using a computer.

What if you use different fonts? Are there any letters that have different types of angles when you use a different font?

Let's Investigate

$6m - \frac{1}{2}m - 30h$

Look at the formula above: *h* stands for the hour and *m* stands for the number of minutes past the hour.

This formula can be used to work out the angle between the hands on an analogue clock.

- If your answer is negative, ignore the negative sign.
- If the hour is 12, use 0 (it is in the zero position).

Use the formula to check your answers to the Let's Investigate activity at the bottom of page 3.

Now use this formula to work out the angles between the hands on an analogue clock for various times that are not 15, 30 or 45 minutes past the hour.

Let's Investigate

Look around the school and find as many different angles as you can.

Sort the angles you find into the following categories:

- Right angles
- Acute angles
- Obtuse angles
- Reflex angles

Hint! Look at:

- Toys
- Objects with moving parts
- Gym equipment

S&C Volume 5

The Maths Herald

Name: Date:

Let's Investigate

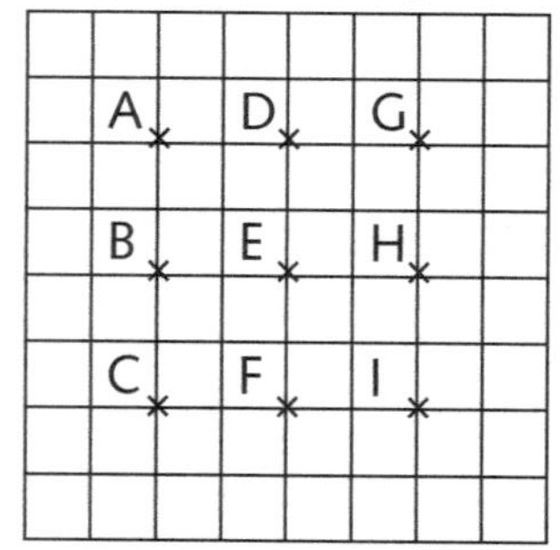

You can travel along the grid lines from Point A to Point E in these ways:

- Down 2, right 2.
- Right 2, down 2.
- Right 1, down 2, right 1.
- Down 1, right 1, down 1, right 1.

How else can you travel along the grid lines from Point A to Point E?

What is the shortest route you can travel between Points A and E?

Choose pairs of points on the grid and investigate different possible routes.

Construct

Each pair of degrees below are 2 of the angles in a triangle.

Work out the size of the missing angle (*x*) in each of these triangles and use a protractor to draw the triangles. Be sure to label the size of each of the angles.

60°, 80° and *x*°

90°, 40° and *x*°

30°, 80° and *x*°

40°, 40° and *x*°

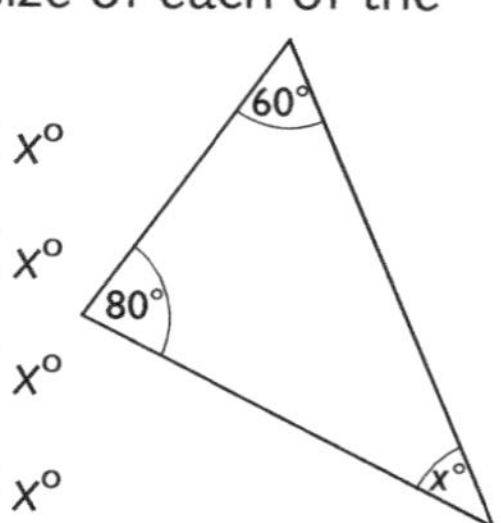

Let's Investigate

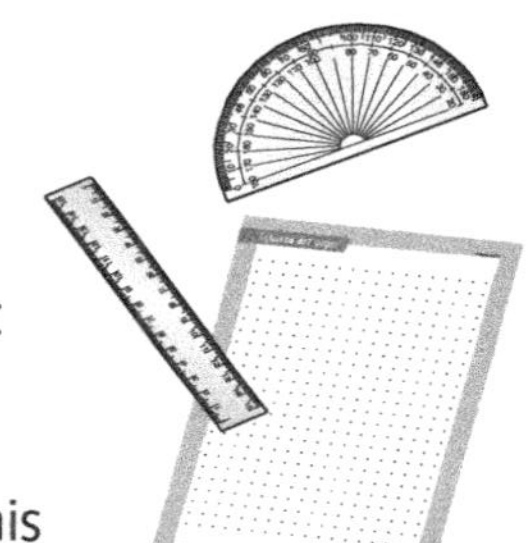

Draw a triangle on squared dot paper.

Using a protractor, measure the size of the 3 angles and mark them on the triangle.

Now draw a parallelogram using this triangle and measure and mark the angles.

What do you notice?

Now draw the same triangle again and make a trapezium using this triangle. Measure and mark the angles of the trapezium. What do you notice?

Let's Investigate

Look at the diagram on the right.
Two of the shortest ways of moving from Point A to Point B are:

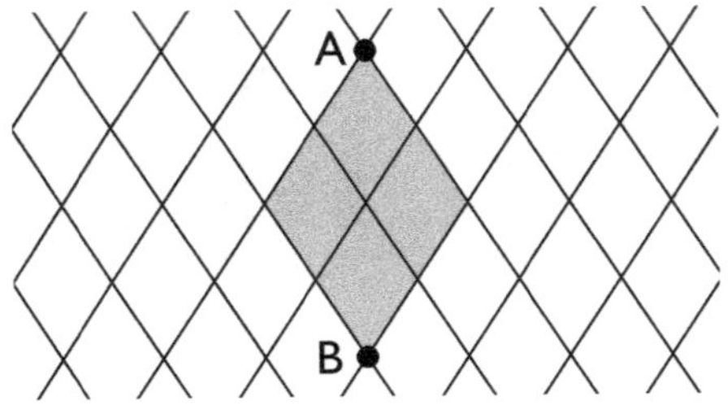

Down left 2, down right 2.

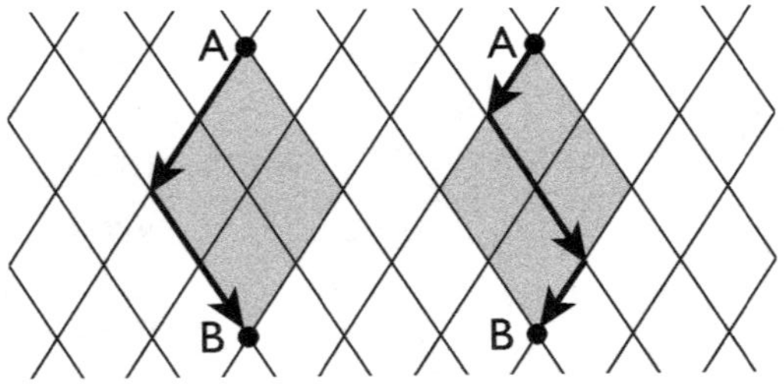

Down left 1, down right 2, down left 1.

What other ways can you find for moving from Point A to Point B that are as short?

Use the diamond paper to show your different routes.

What about for these diagrams on the right?

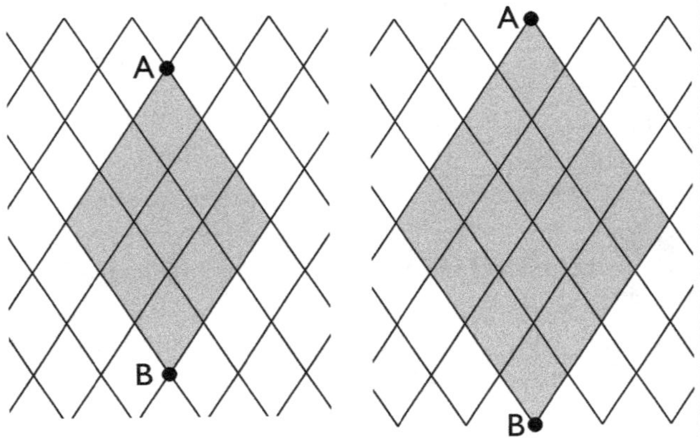

Let's Investigate

What time is it now?

Using a protractor, measure the approximate angle formed by the hour and minute hands of the clock.

Make up some different times and find the angles.

What about making up some angles, e.g. 42°, 65° and 112°, and showing possible times on an analogue clock? Use Resource sheet 9 to record your work.

Let's Investigate

The sum of the angles of a triangle is 180°.

Here's a way to prove it.

Draw a triangle.

Label the angles A, B and C.

Cut out the triangle.

Tear the triangle into 3 pieces, each with an angle.

Place the angles next to each other so that they make a straight line – 180°.

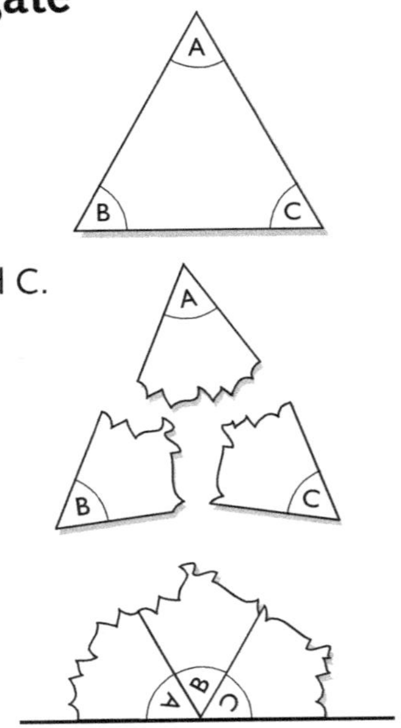

Hint:
You will probably need to draw 1 or more lines to make triangles, for example:

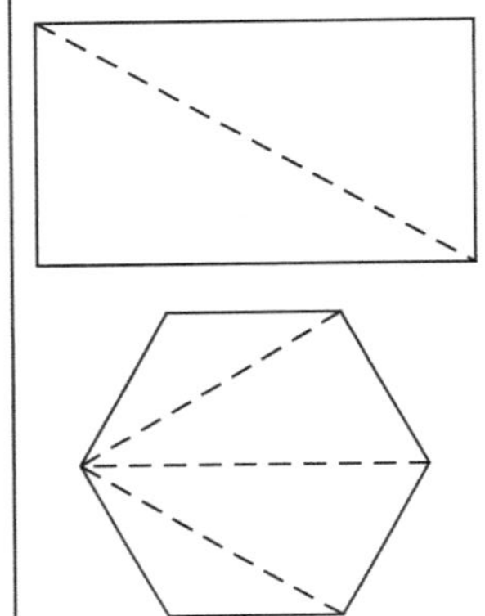

Now use this fact to find the sum of the angles of each of the polygons on Resource sheet 10.

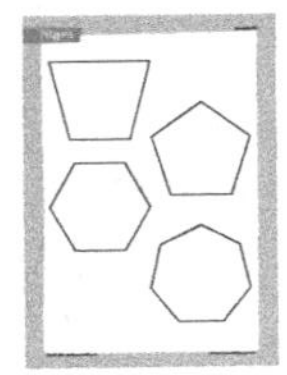

When you have finished, look carefully at your results. What do you notice?

Write a formula for calculating the sum of the angles in a regular polygon.

Let's Investigate

Without using a protractor, what is the angle between the hands on an analogue clock at these times?

- 12 o'clock
- 3 o'clock
- 6 o'clock
- 9 o'clock

How did you work these out?

What are the angles between the hands on a clock at 15 minutes, 30 minutes and 45 minutes past the above times?

How did you work out these angles?

Construct

Some people believe that ink blots are a way to measure how clever and how happy someone is. Herman Rorschach (1884–1922) developed this idea and created the *Rorschach Ink Blot Test*.

What can you see in each of our ink blots?

How many lines of symmetry can you see in each ink blot?

Use ink or paints to make 4 ink blots.

Write about each of your ink blots describing what you can see and its symmetrical properties.

Is it possible to make an ink blot with more than one line of symmetry?

Let's Investigate

Many rugs and carpets show detailed examples of reflective symmetry.

Investigate symmetry in rugs and carpets.

Make a display. Be sure to comment on the symmetry in each of your examples.

S&C Volume 5

The Maths Herald

Name: Date:

Construct

Cross-stitch needlework pictures or designs are made from patterns of 2 diagonal stitches crossing each other.

Using squared paper, design a cross-stitch pattern involving reflective symmetry.

Think about vertical, horizontal and diagonal lines of symmetry.

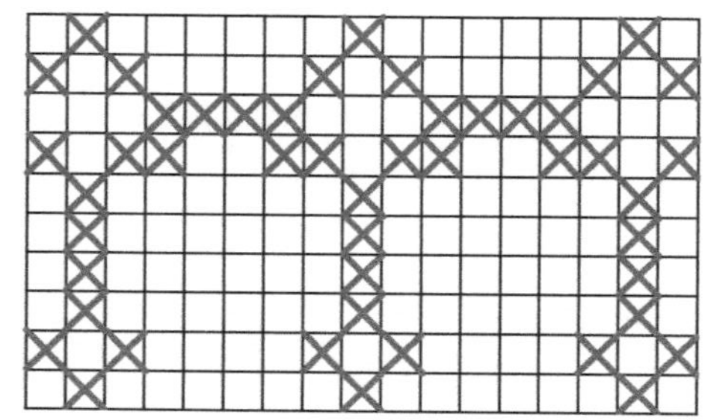

Let's Investigate

A logo is a symbol or design used by an organisation to make it easily recognisable.

Investigate symmetry in logos.

Make a poster showing examples of logos that have symmetry. For each logo briefly describe the type of symmetry that the logo has.

Construct

Your school probably has a logo. It is normally displayed on the notice board at the front of the school and on the school's letterhead.

If your school doesn't have a logo, then design one.

If your school does have a logo, then design a logo for your class.

Whichever logo you design, it must have some form of symmetry.

Remember, a logo needs to be unique to the group of people it is designed for and should be easily recognisable.

Does your school have any sports teams or special interest clubs? If so, design a logo for each of these. If it is a school team or club, you might want to incorporate some element of the school's logo into the design.

Let's Investigate

Using colour, investigate the different patterns you can make from this shape using reflective symmetry.

Record your patterns on Resource sheet 8.

Be sure to draw in the lines of symmetry.

The Puzzler

On a calculator display 2 is the reflection of 5.

Which 3-digit numbers display reflective symmetry about a vertical line through the centre of the digits?

Which 4-digit numbers display reflective symmetry about a vertical line through the middle 2 digits?

Let's Investigate

Using 1 of the grids on Resource sheet 11, reflect Shape A in the x-axis. Then reflect the resulting shape (Shape A^1) in the y-axis. Label the new shape: Shape A^2.

Now, using another grid on Resource sheet 11, reflect Shape A in the y-axis. Then reflect the resulting shape (Shape A^1) in the x-axis. Label the new shape: Shape A^2.

What do you notice about both Shape A^2s?

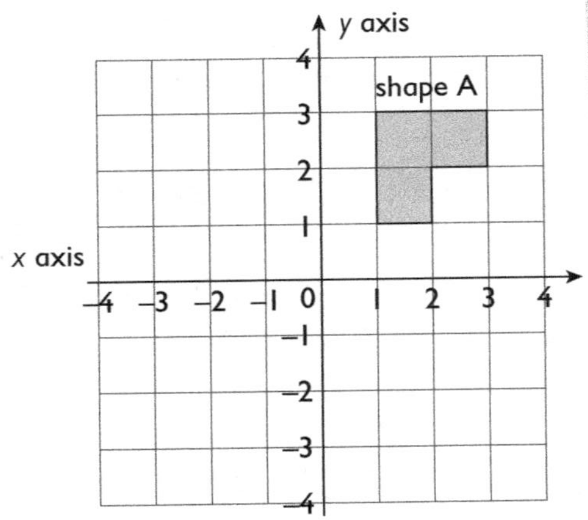

Let's Investigate

Investigate the following generalisations about reflective symmetry.

A reflection in 2 parallel lines is equivalent to a translation.

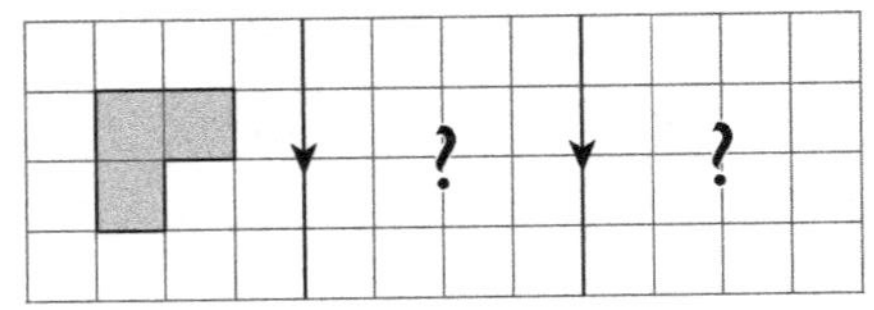

A reflection in 2 perpendicular lines is equivalent to a half-turn rotation.

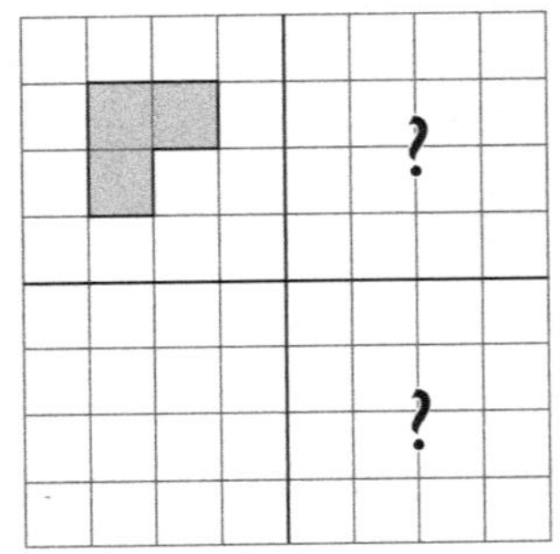

Focus on Science

Look at these illustrations.

They all show examples of reflective symmetry.

How?

Investigate other examples of reflective symmetry in nature.

Around the World

Using 1 cm squared paper, draw a grid and letter 14 horizontal squares A to N and number 12 vertical squares 1 to 12 as in the diagram below.

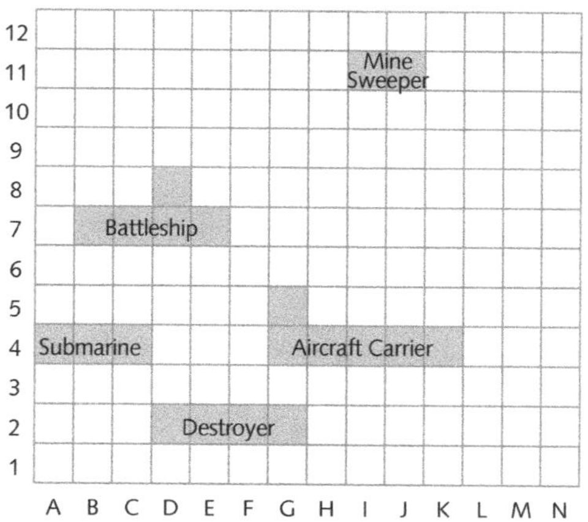

Translate each ship in the diagram according to the following instructions:

Translate the:

- Mine Sweeper: 5 squares to the left and 6 squares down.
- Battleship: 7 squares to the right and 5 squares down.
- Submarine: 9 squares to the right and 5 squares up.
- Aircraft Carrier: 6 squares to the left and 7 squares up.
- Destroyer: 7 squares to the right and 4 squares up.

Now write the grid references of the new positions of each of the ships.

For example, the current position of the Battleship is B7, C7, D7, E7 and D8.

Around the World

The sailing boat occupies squares A5, B5, C5 and B6.

It needs to sail to the shaded squares on the other side of the island.

It can sail 5 squares each day, either up, down, left or right, or any combination in 24 hours.

Draw in the ship's position after each 24 hours and then write down each of the corresponding grid references.

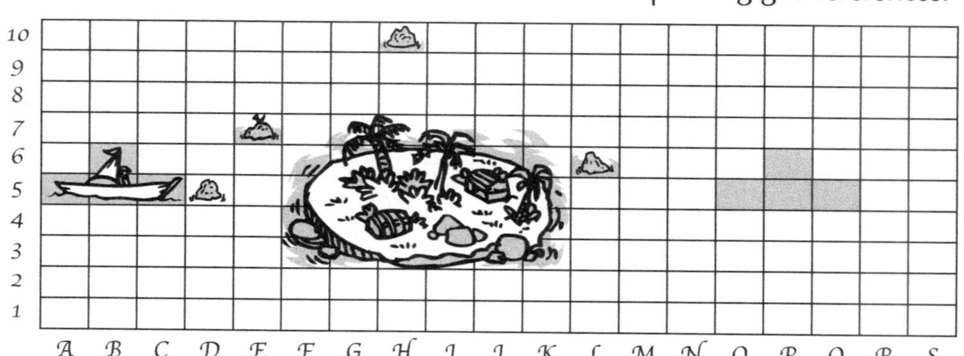

S&C Volume 5

The Maths Herald

Name: Date:

Construct

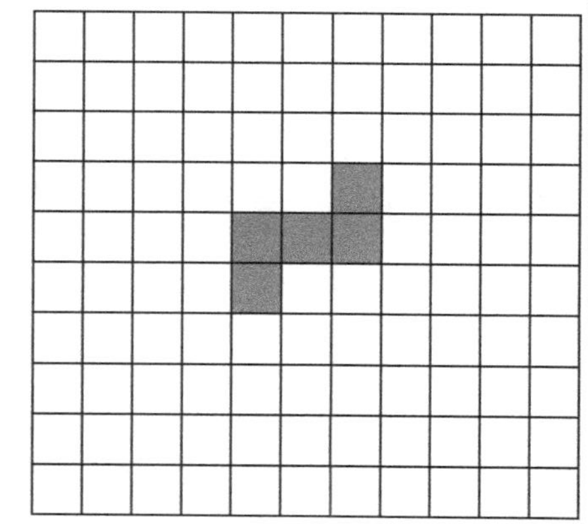

Use a sheet of squared paper. In the centre of the sheet, colour adjoining squares to make a simple, but interesting, shape.

Then using 2 coins of different denominations and a 1–6 dice, follow these rules to make a translating pattern.

Coin A: Heads = Move up

Tails = Move down

Coin B: Heads = Move to the left

Tails = Move to the right

1–6 dice: The number of squares to move.

Investigate making similar patterns using diamond paper.

Looking for Patterns

tessellation

Cut out the large square on Resource sheet 12 to make 4 smaller squares.

Using all of the 4 small squares, arrange them to make 1 large square pattern.

For example,

Repeat your pattern on squared paper to make a tessellating pattern.

Change the position of the 4 small squares to make a different pattern and repeat.

How many different patterns can you make?

Construct

Cross-stitch needlework pictures or designs are made from patterns of 2 diagonal stitches crossing each other.

Using squared paper, design a cross-stitch pattern involving translation.

Think about vertical and horizontal translations.

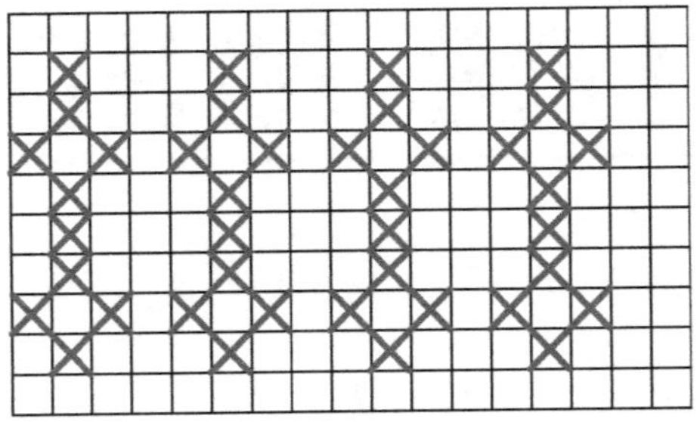

The Language of Maths

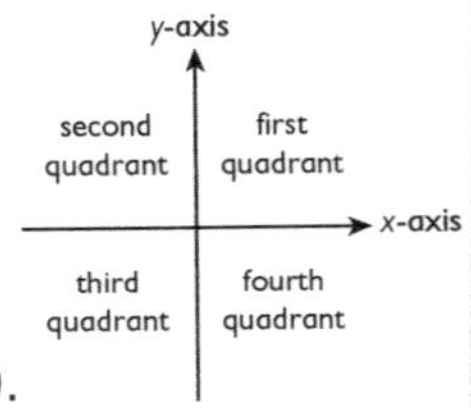

This diagram shows the 4 quadrants of a coordinates grid.

The coordinates below are for Shape A^1 and Shape A^2.

(Shape A^2 is a translation of Shape A^1 in the second quadrant).

Shape A^1

(3,1), (3,2), (4,2), (4,3), (1,3), (1,2), (2,2), (2,1), (3,1)

Shape A^2

(–2,1), (–2,2), (–1,2), (–1,3), (–4,3), (–4,2), (–3,2), (–3,1), (–2,1)

Translate Shape A^1 and Shape A^2 4 squares down and draw them on Resource sheet 11.

What are the coordinates for these shapes?

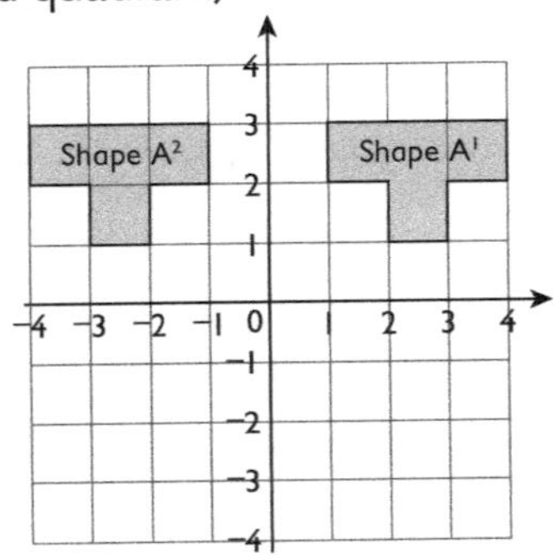

In the Past

William Morris (1834–1896) was a craftsman born in East London. He is best known for his repeating patterned fabric and wallpaper designs, many of which are based on nature.

Investigate the designs of William Morris, commenting on any translations.

What about other transformations? www

Looking for Patterns

Shape A has been translated a total of 10 times.

We can say that Shape A^1 has been translated 4 squares up and 2 squares to the left, or 2 squares to the left and 4 squares up.

Write different ways to show how shapes A^1 to A^{10} have been translated.

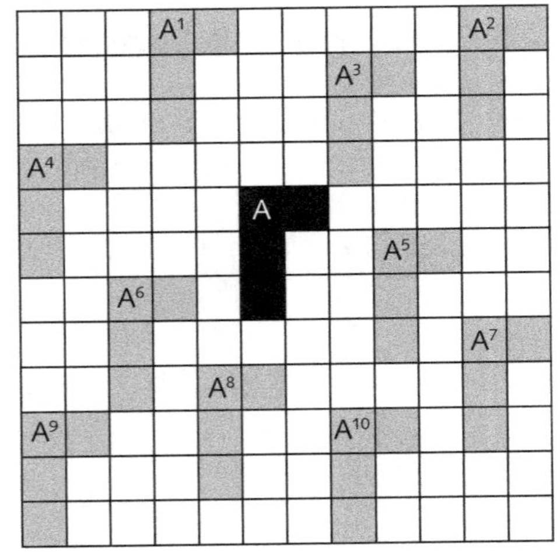

The Arts Roundup

Piet Mondrian (1872–1944) was an abstract artist known for his paintings that used large vertical and horizontal lines, and rectangles of primary colours. The examples below are drawn in his style.

Investigate the work of Piet Mondrian.

Design your own Piet Mondrian inspired work. Ensure that you include some translations. www

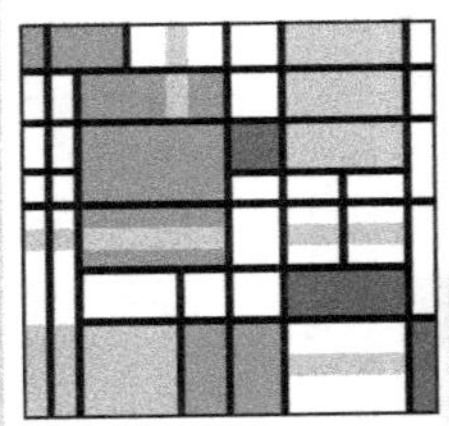

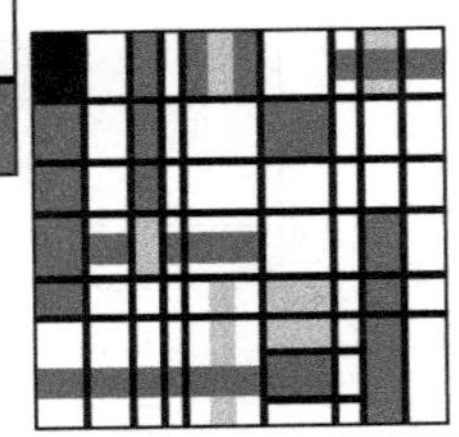

Around the World

Aboriginal art is art made by the Indigenous Australians. The Australian Aboriginal culture is one of the most ancient in the world and its art is largely based on mythology referred to as *Dreamtime*. The work is characterised by hundreds of dots surrounding shapes with a symbolic meaning. The symbols are used to create a simple storyline.

Investigate Aboriginal art.

Design your own Aboriginal art inspired work. Ensure that you include some translations. www

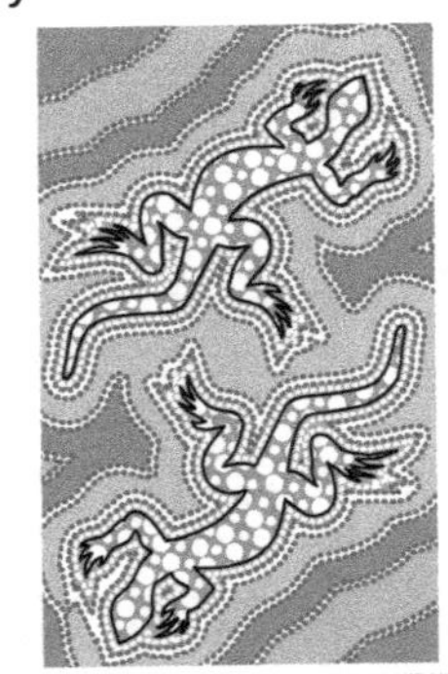

Let's Investigate

You can find examples of rotational symmetry in kitchens and bathrooms.

Plug holes have rotational symmetry.

Look at some plug holes in your home and at school.

Either draw a sketch of them or take a photo.

Make a display. Be sure to comment on the symmetry in each of your examples.

The Language of Maths

Investigate all the letters of the alphabet for rotational symmetry.

Try for both UPPER CASE and lower case letters. For example,

Aa Bb Cc …

The Language of Maths

Playing cards show rotational symmetry.

Why do you think this is?

Find other examples of cards with rotational symmetry.

We often use 0–9 digit cards during mathematics lessons.

Design a set of 0–9 digit cards showing rotational symmetry.

S&C Volume 5

The Maths Herald

Name: Date:

Construct

Cross-stitch needlework pictures or designs are made from patterns of 2 diagonal stitches crossing each other.

Using squared paper, design a cross-stitch pattern involving rotational symmetry.

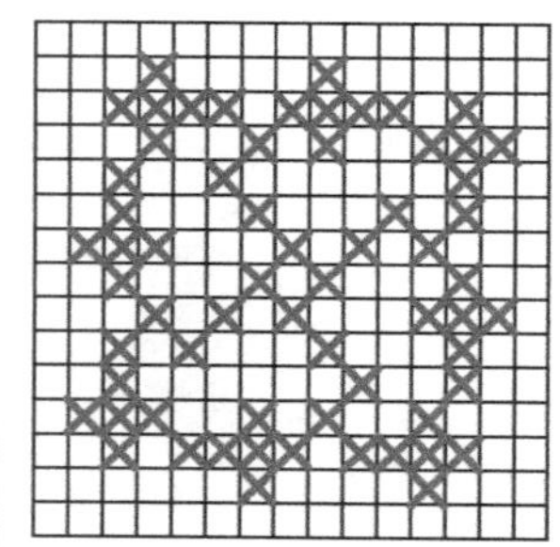

Technology Today

Rotate each of these shapes clockwise 90°. Keep doing this.

Show where the star is on each new rotation.

Stop when you have rotated each shape one complete revolution.

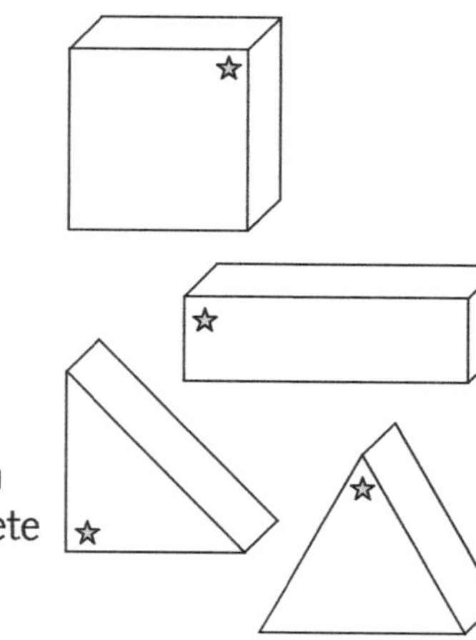

How are you going to show your results?

Can you show them with the help of a computer?

The Language of Maths

A *kaleidoscope* is a toy consisting of a cylinder with mirrors and small coloured shapes made from glass, plastic or paper. The shapes produce changing symmetrical patterns when the cylinder is rotated.

Find images on the Internet of the patterns made by kaleidoscopes.

Make a display. Be sure to comment on the symmetry in each of your examples.

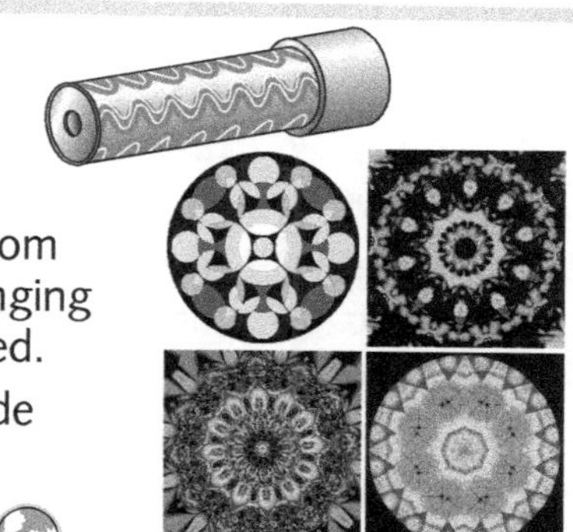

Let's Investigate

Using colour, investigate the different patterns you can make from this shape using rotational symmetry.

Record your patterns on Resource sheet 8.

Around the World

This tiling pattern can be found in the Alhambra Palace in Granada, Spain.

Write a description of this pattern using the language of rotational symmetry.

Around the World

The Alhambra is a palace and fortress complex constructed during the mid 14th century by the Moorish rulers of Granada, Spain.

It is now one of Spain's most popular tourist attractions, containing the country's most significant and well known Islamic architecture, along with 16th century and later Christian buildings and gardens.

The Alhambra is a UNESCO World Heritage Site.

The Alhambra has many fine examples of reflective and rotational symmetry.

Find out about the symmetrical patterns found in The Alhambra and make a display commenting on the symmetry contained in each pattern.

Looking for Patterns

The pattern on the right has been made using this tile.

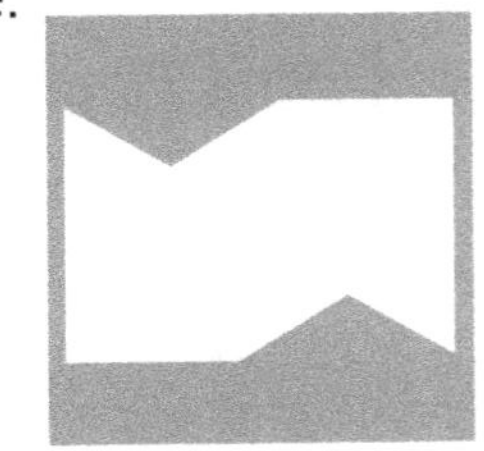

Make a tile of your own and design your own rotational pattern.

Looking for Patterns

Imagine rotating a right-angled triangle through 180° about the mid-point of its shortest side. Draw the shape.

What shape have you drawn?

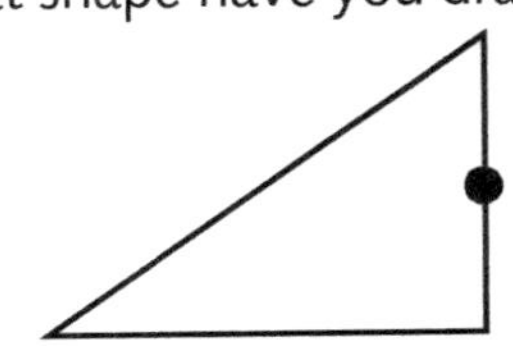

What happens when you rotate the triangle about the mid-point of its longest side? Draw the shape.

What shape have you drawn?

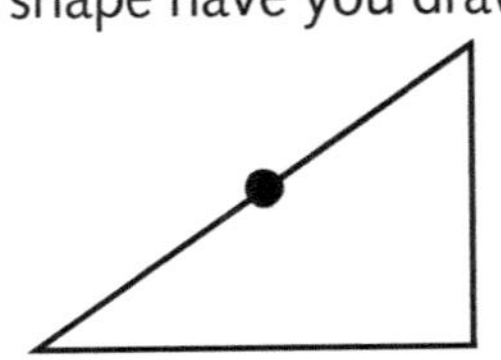

Focus on Science

Snowflakes are an example of both reflective and rotational symmetry in nature.

Using paper and scissors, and by folding, make some symmetrical snowflakes.

Try and make your snowflakes as detailed as possible.

Write a description of each of your snowflakes describing their symmetrical properties.

Around the World

The compass rose in the Looking for Patterns activity on page 3 is an 8-point compass rose.

The compass rose on the right is a 16-point compass rose.

Complete the compass rose by writing the abbreviations for the missing directions.

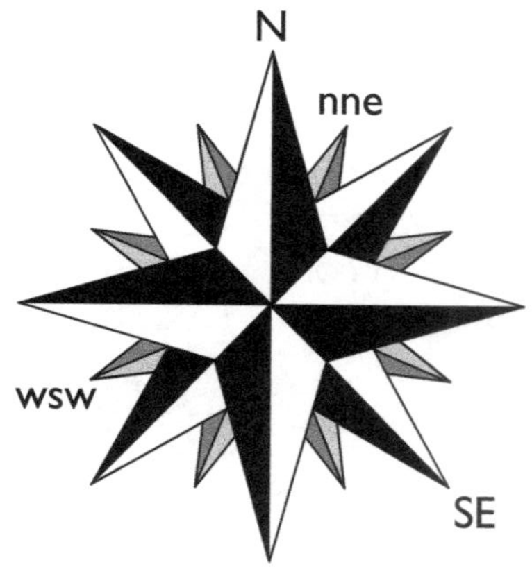

Looking for Patterns

Referring to the 16-point compass rose in the Around the World activity above, if you move in a clockwise direction, the size of the angle between N and nne is 22·5°.

What is the size of the angles between these points if you move in a clockwise direction around a 16-point compass rose?

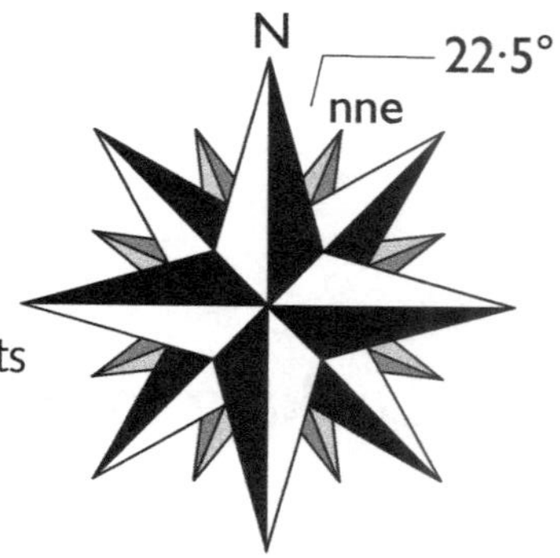

- nne and wnw
- ssw and sse
- ese and sse
- N and wsw
- E and ssw
- ene and wsw
- nnw and NW
- wnw and nne
- W and ene
- SE and ssw
- S and NE
- ese and NE
- ssw and nnw
- N and ssw

Look at your answers. What patterns do you notice?

S&C Volume 5

The Maths Herald

Name: Date:

Around the World

A new housing estate is being built.

The Post Office need to decide the best location for a post box.

Where should the PO locate the post box?

Justify your decision.

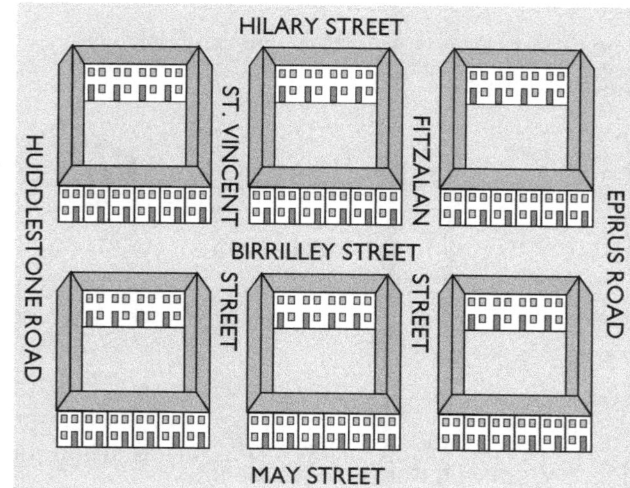

Around the World

The Post Office needs to plan the most efficient route for a post-person to deliver the mail on the new housing estate above.

If the post-person arrives and leaves the estate at the corner of May Street and Epirus Road, what would be the shortest route possible?

Is there more than 1 possibility?

At Home

Write a set of instructions for someone to follow that describes, in detail, your route from home to school.

Then draw them a map to follow.

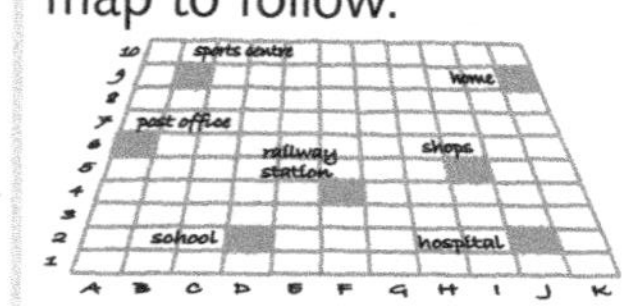

Construct

The points (–2,1) and (3,1) are 2 vertices of the 4 vertices of a rectangle.

What are the coordinates of the other 2 vertices?

What are all the different possibilities?

Show them on Resource sheet 11.

Then calculate the perimeter and area of each rectangle.

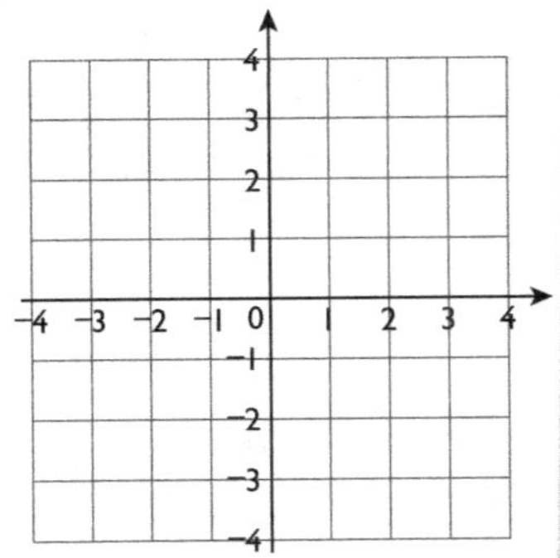

Let's Investigate

Plot these 3 points on one of the coordinates grids on Resource sheet 11.

(1,2), (–2,1), (–1,3).

Write the coordinates for the fourth point that will make a kite. Join the coordinates to draw the kite.

What other quadrilaterals can you draw with 3 of the shape's vertices at points (1,2), (–2,1), (–1,3)?

What quadrilaterals can you not draw using these three vertices? Why?

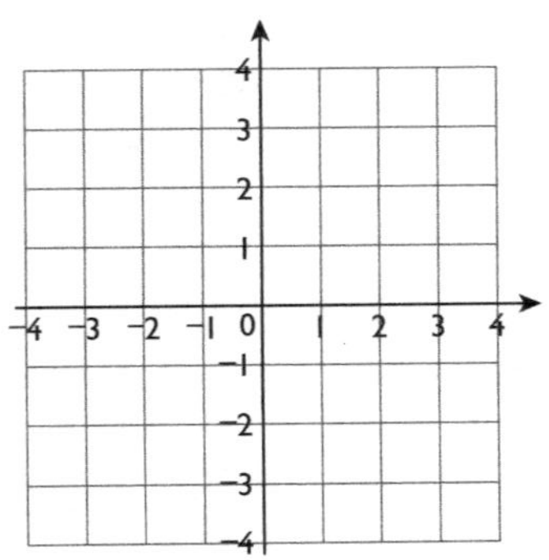

Construct

Using squared paper, draw a map of Laura's walk to school.

- Laura walks to school with her mum in the morning.
- They turn right out of their front gate.
- They turn left at the post box.
- They turn into the next street on the right, by the large tree.
- At the T-junction, in front of the library, they turn right.
- At the traffic lights they turn left.
- They then turn into the first road on the left.
- The school is on the right-hand side of the road.

Describe Laura's walk home from school in the afternoon.

Around the World

Using a street directory or local map, find the grid reference for each of the following:

- your school
- your home
- the nearest set of shops
- a local sports facility
- the nearest railway station
- the nearest post office
- the nearest hospital

Around the World

Using the information you have gathered in the Around the World activity on the left, draw a map showing the location of each of the 7 features.

Be sure to include your own grid referencing system.

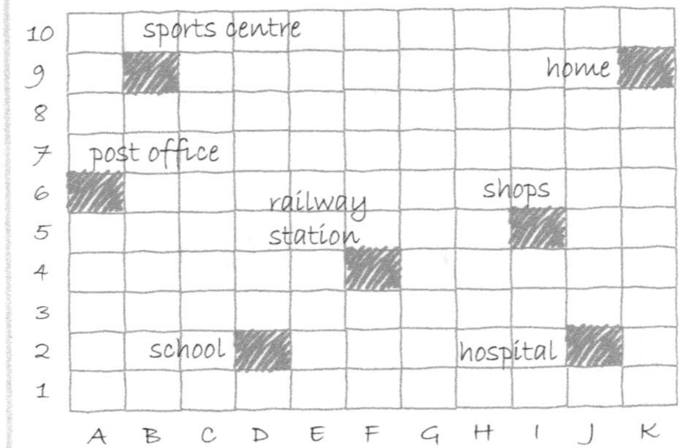

Looking for Patterns

If you move in a clockwise direction, what is the size of the angle between these points?

- E and S
- N and S
- N and NE
- SW and NW
- E and NW
- W and SW
- SE and N
- NE and S
- N and NW
- W and E
- SW and W
- N and SE

Look at your answers.

What conclusions can you make about the size of the angles when moving between different directions on an 8-point compass rose?

Construct

Origami is the Japanese art of paper folding.
Follow these instructions to make an origami star.

You will need 6 paper squares, ideally each a different colour.

Step 1:
Take 1 piece of paper and fold it in half and half again. Crease, then open out the square.

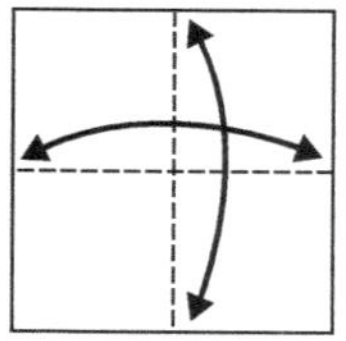

Step 2:
Turn the paper over and fold it in half diagonally in both directions. Crease, then open out the square again.

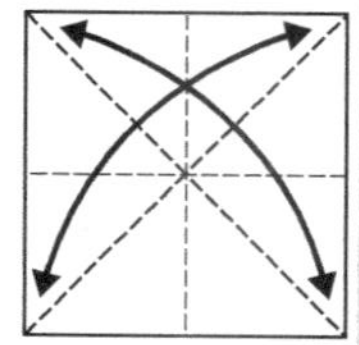

Step 3:
Holding the 2 points on either side of the square, bring them both down to the centre point on the bottom side. Crease, then flatten the model.

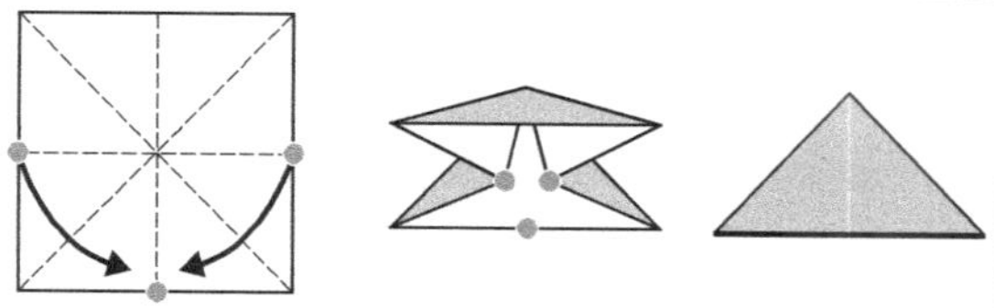

Step 4:
Fold the bottom corners of the top layer only to the top point then unfold. Turn over and do the same. This is 1 origami piece.

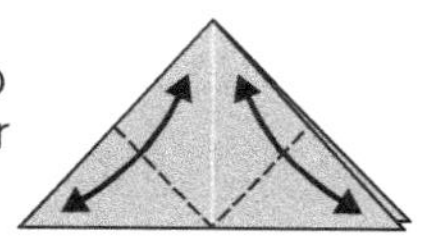

Step 5:
Repeat Steps 1–4 to make another 5 origami pieces, giving you 6 altogether.

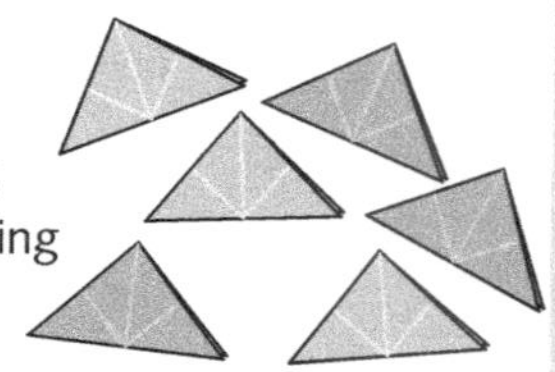

Step 6:
To join the origami pieces together, slip 1 set of 'arms' into another set of arms, as shown.

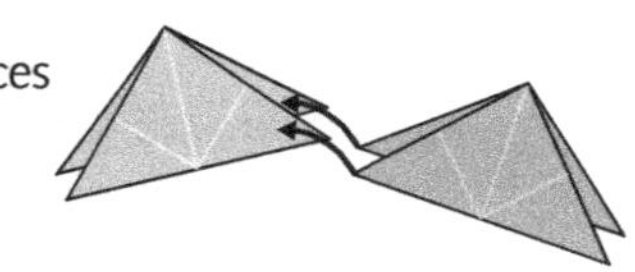

Step 7:
Push the arms in as far as they will go.

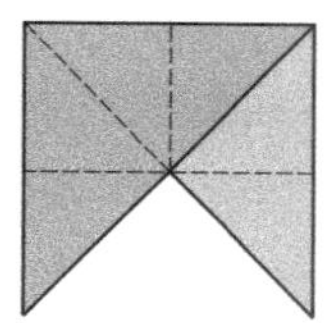

Step 8:
Pull the 2 top creases apart.

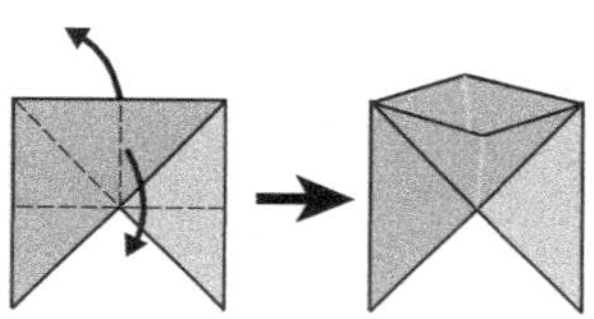

Step 9:
Repeat Steps 6–8 to add the origami pieces to build the star.

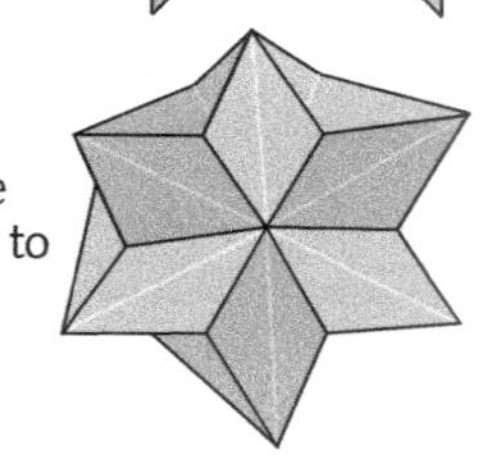

The Language of Maths

What is the difference
between a square,
a rectangle and an oblong?

S&C Volume 5

The Maths Herald

Name: Date:

Around the World

In many countries in parts of Eastern Europe, it is customary to decorate eggs at Easter with elaborate geometric designs.

Follow the steps below to make an Easter egg design.

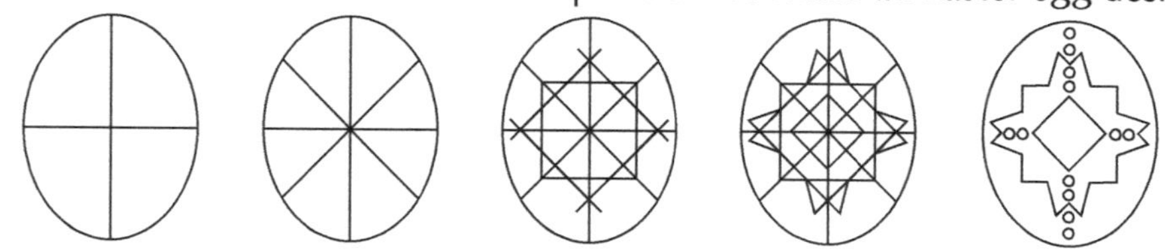

Using Resource sheet 13, make an Easter egg design of your own.

The Puzzler

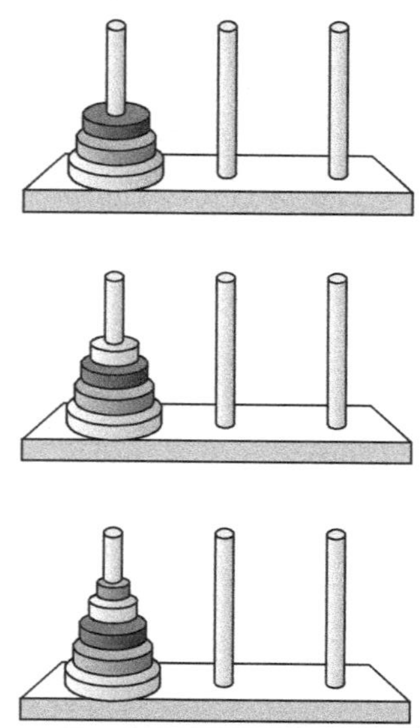

The Tower of Hanoi is a puzzle that consists of 3 rods, and a number of disks of different sizes that can slide onto any rod.

The puzzle starts with the disks in a stack in order of size on 1 rod, the smallest at the top.

The objective of the puzzle is to move the entire stack to another rod.

These are the rules:

- Only 1 disk can be moved at a time.
- Each move consists of taking the top disk from one of the rods and putting it onto another rod, on top of the other disks that may already be on that rod.
- Do not place a bigger disk on top of a smaller disk.

Start with 3 disks, then move on to using 4 disks and finally 5 disks.

Looking for Patterns

Use Resource sheet 14.
For each of the grids draw straight lines between points to make different patterns.
What is your rule for each pattern?

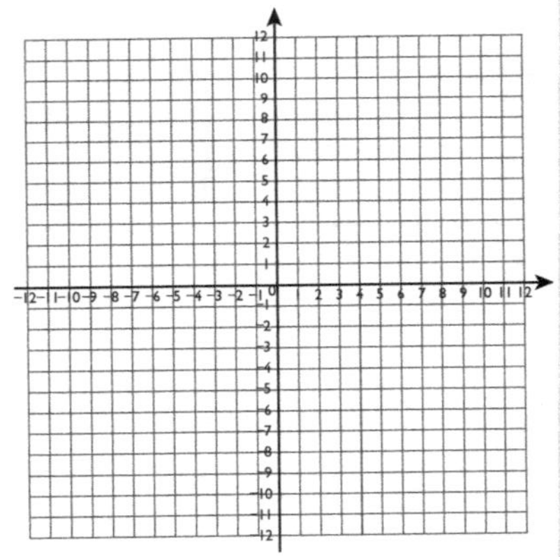

What patterns can you make using the grids on Resource sheets 15, 16 and 17?

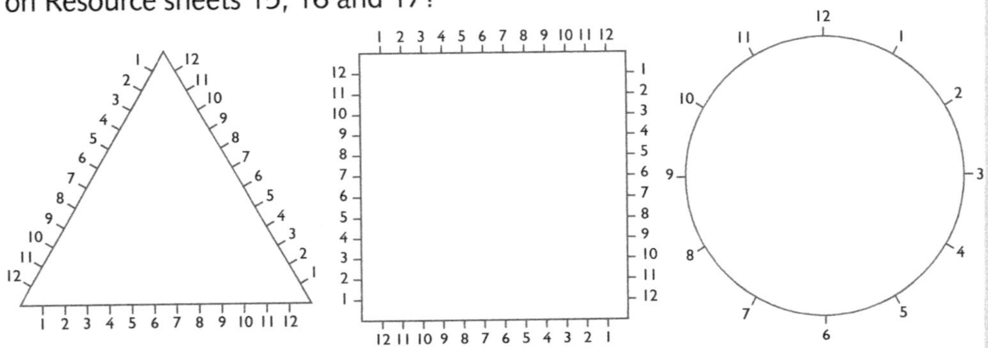

The Puzzler

Draw a large 3 × 3 grid.
Place 8 interlocking cubes as shown:
7 cubes of 1 colour, 1 cube in a different colour.

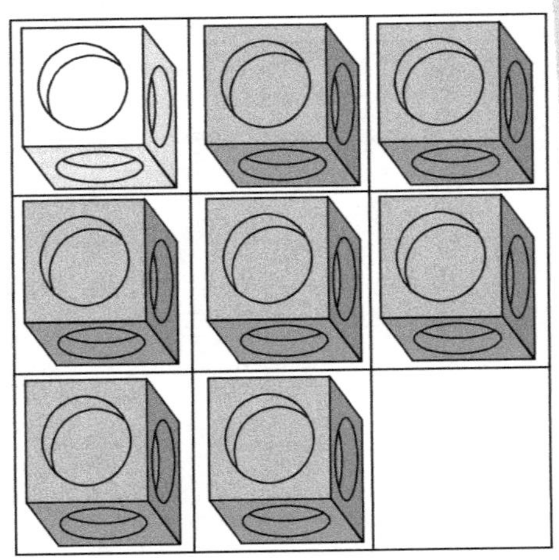

By moving one cube at a time around the grid, can you move the white cube into the bottom right-hand corner?

What is the shortest route you can find?

If that's too easy, what about trying 15 cubes on a 4 × 4 grid?

Looking for Patterns

The word *quincunx* is used to describe an arrangement of 5 objects in a square, with 4 at the corners and 1 in the centre.
What objects can you think of that are arranged in quincunx formation?

Design your own quincunx square.
Use this as a template to make a repeating quincunx pattern.

You may want to use colour and an interesting object to make your pattern more exciting.

Let's Investigate

Cross-stitch needlework pictures or designs are made from patterns of 2 diagonal stitches crossing each other.
Make a display of different cross-stitch designs.
For each design state what form of transformation the design has:

- Reflective symmetry
- Rotational symmetry
- Translation

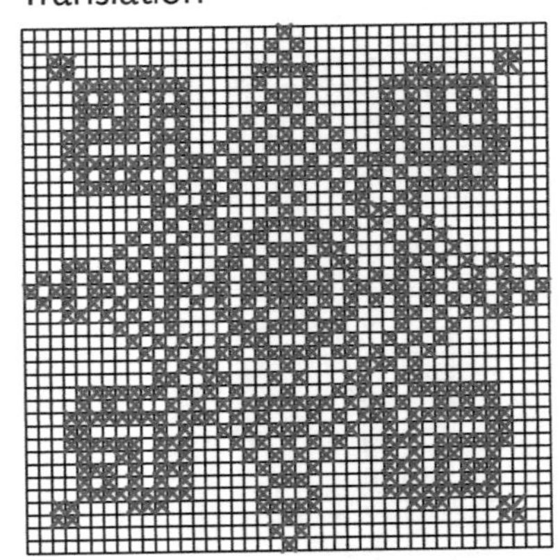

Let's Investigate

If 2 or more 2-D shapes have the same shape and size, and with corresponding sides and angles that are equal, then the shapes are said to be *congruent*.
How many different ways can you find of dividing a 4 × 4 geoboard into 2 congruent halves?
Here are 2.

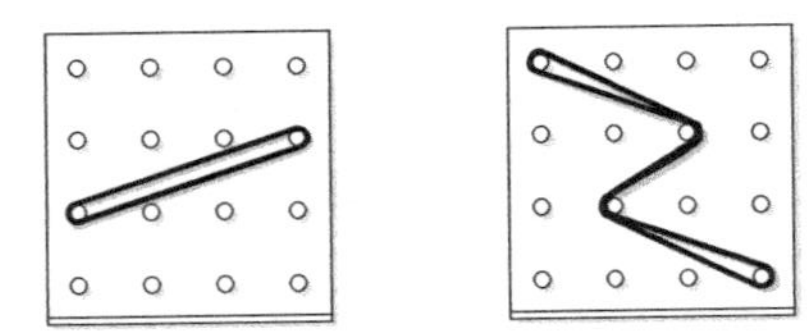

Beware! These 2 are considered to be the same.

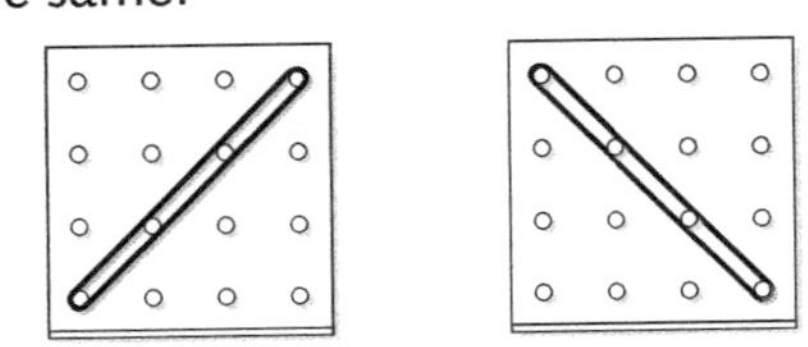

The Arts Roundup

Look at this graph. It shows the number of cinema tickets that were sold during 1 week for *The Return of the Phantom X*.

This data has been displayed as a bar chart.

How else could you represent this data?

Represent the data in as many different ways as you can.

Write a statement about the advantages and disadvantages of each type of representation.

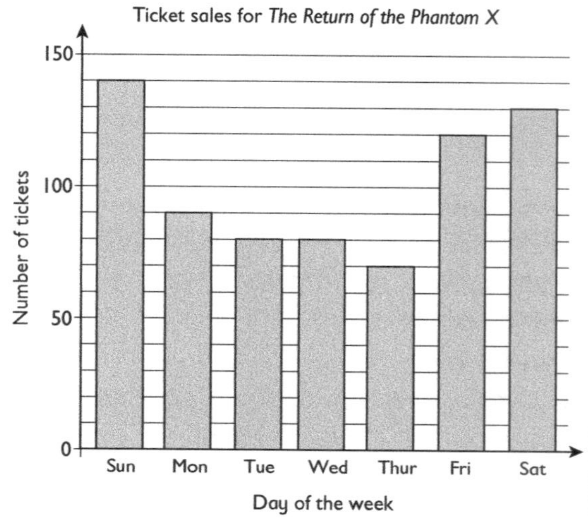

The Arts Roundup

Look at the bar chart in The Arts Roundup activity above.

Write a detailed report explaining the data.

Make inferences about the number of tickets sold on each particular day.

Be sure to justify your inferences.

S&C Volume 5

The Maths Herald

Name: Date:

Around the World

By sea, what are the different routes for crossing the English Channel from the UK?

How much does each one cost?
Does the cost vary throughout the year?

How long does each one take?

Which route would you use and why?

Let's Investigate

Investigate the statement: *Taller people have bigger feet.*

List the information you need to collect.

Collect the information of 10 people and record it in a table.

What if you asked 10 other people? Write about what might happen.

The Language of Maths

Look at this bar chart.

What might the data in the graph represent?

Label the vertical and horizontal axes and write a title for the graph.

Write an explanation as to why your labels are appropriate for this set of data.

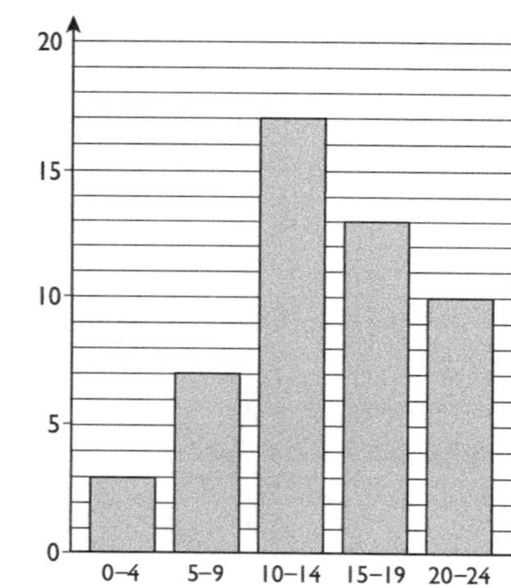

Famous Mathematicians

Lewis Carroll, who wrote *Alice's Adventures in Wonderland* and *Alice Through the Looking Glass*, was not only a famous children's author but also a great mathematician. His real name was Charles Dodgson and he was born in England in 1832.

Dodgson was fascinated with solving mathematical problems, especially *logic puzzles*. To help him with organising his thoughts when solving such problems he invented the *Carroll diagram*.

	1-digit numbers	Not 1-digit numbers
Even	2, 4, 6, 8	106, 78, 10, 36, 12
Not even	3, 5, 7, 1, 9	11, 93

Dodgson also created a type of puzzle known as a *doublet*. To solve a doublet you have to change 1 word into another word. You do this by changing it 1 letter at a time into a new word.

For example, to change the word 'more' into the word 'less':

more
lore
lose
loss
less

Solve these doublets.

- Increase COLD to WARM
- Make FIRE produce HEAT
- Change HAND to FOOT

Create some doublets for friends to try and solve.

Looking for Patterns

- Draw a simple picture on one side of a piece of paper.
- Hold the paper as high as you can with the picture facing upwards.
- Drop it.
- Record the side it lands on (picture or blank).
- Do this 20 times.

Write about the results.

Drop the paper another 20 times. Write about the results.

Compare them with the results of the previous 20 drops.

Do it again another 20 times. Write about the results.

Compare this set of 20 results with both the previous 20 drops.

Combine the 3 lots of 20 drops for a total of 60 drops.

Write about what you notice.

Sports Update

Many sports and games involve an element of skill. The better your skill, the better chance you have of winning. The level of skill can depend on many factors including natural ability and the amount of practice.

Design an experiment to test whether or not your level of skill improves with practice. For example, it might be an activity that involves throwing beanbags into a bucket.

Think about:

- What you will need to measure at the start of the experiment.
- What constitutes 'enough' practice.
- How you will measure improvement.
- How you will measure whether or not the amount of practice you have had has increased your level of skill.

Has your chance of winning improved? By how much?

Sports Update

Work with a partner to find out if Tina is correct.

- Look through the television schedules of sports programmes for one day.
- Make a list of all the sports.
- Record how many times each sport is shown.
- Which sports are on most often? Make a list of the top 5.
- Work out approximately how many hours of each sport are shown.

Is Tina correct?

Write a report of your findings. Find ways to present your information clearly.

What if you looked at the television listings for a different day?

Around the World

These climate graphs show the average annual rainfall and temperature for four different cities.

Imagine you work for a travel agency. Write a report of each city recommending which months of the year are best for tourists to visit each one, and which months they should avoid.

Provide a rationale for each of your recommendations.

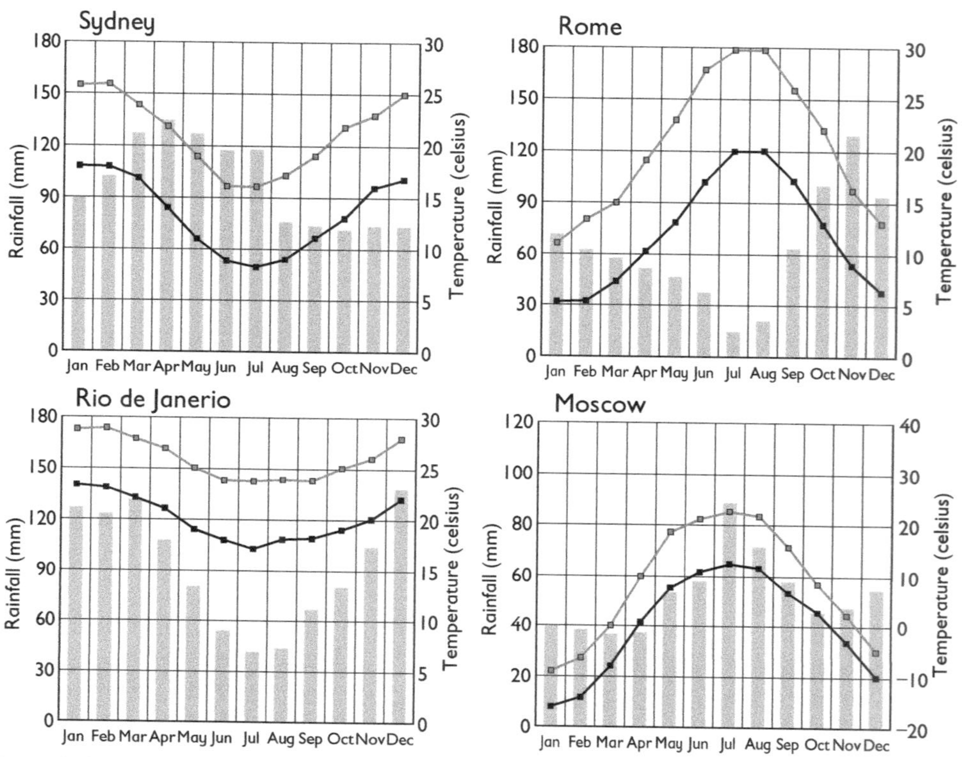

S&C Volume 5

The Maths Herald

Name: Date:

Sports Update

Draw a graph that shows your change of pulse rate after exercise.

First take your pulse rate for 1 minute while at rest.

Then do some strenuous exercise, such as running, for 5 minutes.

As soon as you have finished, take your pulse rate again for 1 minute.

Then, every 5 minutes, keep taking your pulse rate until it is the same as it was at rest.

Then draw the graph.

Let's Investigate

Carry out a survey of the viewing habits of the families of children in your class.

You need to find out what are the 10 most popular television programmes amongst all the families.

Consider these factors:

- The different television programmes families watch in a week.
- What proportion of the total amount of viewing time is spent watching each programme.
- The number of people in each family that watch each particular programme.

Follow these steps:

Step 1: Plan
Step 2: Collect data
Step 3: Process the data
Step 4: Represent the data
Step 5: Interpret and discuss the data

At Home

Investigate further the television viewing habits in your home.

Part I

How many different television channels are available to watch in your home?

Of all of these channels, what are the 5 channels that are watched the most?

Part II

Work out the approximate number of television viewing hours at your home in one week.

Of the total weekly television viewing hours, what proportion is spent watching the top 5 channels?

Let's Investigate

Collect several different daily and weekend newspapers.

Make a list of all the different types of articles.

For example:

- News
- Lifestyle
- Editorial
- Special Features
- Classifieds
- Sport

Do different newspapers devote different proportions of the newspaper to each of the article types on your list?

Do these differ for weekday and weekend newspapers?

Let's Investigate

Collect about 6 different cereal boxes.

Use a data collection sheet to collect information from each cereal box.

Draw graphs to represent the:

- fat content
- protein
- calorie value
- carbohydrate content.

Decide which is the healthiest cereal for:

- your heart (low fat)
- energy (calories)
- stamina (carbohydrate)
- protein.

Write a short report of your investigation.

Sports Update

This chart shows the costs at Fast-track Fitness Centre.

Which option is best for someone who visits the fitness centre 7 days a week, 5 days a week, twice a week or once a week?

Fast-track Fitness Centre
Pay per visit: **£5**
Monthly subscription: **£40**

Let's Investigate

Work with a partner.

Which do you think are the busiest areas in your school playground?

Why do you think this is? Do you think it is different for infants and junior children?

Decide how to collect evidence to show that you are right (or wrong)!

Design a data collection sheet.

At a playtime, collect and record your information.

Write a short report about your investigation.

Were you right or wrong about the busiest parts of the playground?

What if you wanted to find out why? How could you do this?

The Language of Maths

Who, in order, are the top 5 favourite authors of all the children in Years 5 and 6?

Follow these steps:

Step 1: Plan

Step 2: Collect data

Step 3: Process the data

Step 4: Represent the data

Step 5: Interpret and discuss the data

Teacher's notes

Issue 1

Number

Prerequisites for learning

- Recognise and use square numbers and multiples
- Explore number sequences, patterns and relationships
- Use simple expressions and formulae in words and symbols

Resources

pencil and paper
Resource sheet 2: My notes (optional)
Resource sheet 3: Pupil self assessment booklet (optional)
calculator (optional)
set of 1–9 digit cards
set of 1–16 number cards (optional)
computer with Internet access

Teaching support

Page 1

Famous Mathematicians

- Ask the children to test the theorem for numbers larger than 400. For this they will need to know the squares of numbers larger than 20.

Let's Investigate

- Work through an example with the children, e.g.
  $3^2 = 9$ and $5^2 = 25$
  $9 + 25 = 34$
  $34 \times 2 = 68$
  $68 = 64$ and 4 ($8^2 + 2^2$)

Looking for Patterns

- Two different rules create this pattern. Can the children identify them both?
- Assist the children in identifying the rules.
- Children investigate patterns made from other rules.

Page 2

Looking for Patterns

- Ensure children can identify the pattern, in particular how to arrive at the last number in each row, i.e.

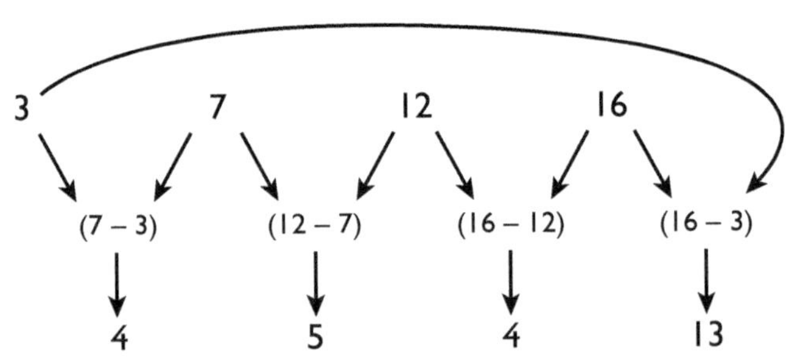

- Children investigate sequences of five, six or seven numbers.

Looking for Patterns

- Suggest the children express their rule using symbols, e.g.
  – If the number is even, halve it.
  If n is even then $\frac{n}{2}$
  – If the number is odd, multiply it by 3 and subtract 1.
  If n is odd then $3 \times n - 1$ or $3n - 1$.

Technology Today

- Ask the children to investigate only one of the systems.
- Ask the children to suggest and investigate their own numbering system, e.g. flight numbers.

Page 3

The Puzzler

- Children use a set of 1–16 number cards arranged in a 4 × 4 grid formation.

Sports Update

- Numbers 1–99 need 20 ones (or twos or threes etc.)
  Numbers 100–199 need 120 ones
  Numbers 200–999 need 8 × 20 ones
  Total number of ones is 20 + 120 + (8 × 20) + 1 = 301
  (1 for 1000).
- Numbers 1–99 need nine zeros
  Numbers 100–999 need 9 × 20 zeros
  Total number of zeros is 9 + (9 × 20) + 3 = 192
  (3 for 1000).
- Total number of digits needed = (300 × 9) + 1 + 192 = 2893.
- An alternate method to working out how many digits there are altogether in the numbers 1 to 1000 is:
  9 × one-digit numbers = 9
  90 × two-digit numbers = 180
  900 × three-digit numbers = 2700
  1 × four-digit number = 4
  9 + 180 + 2700 + 4 = 2893.

Page 4

The Language of Maths

- Children investigate the meanings of the following abbreviations:
  - HCF (Highest Common Factor)
  - GCF (Greatest Common Factor)
  - GCD (Greatest Common Divisor).

The Puzzler

- Suggest children start with the centre hexagon.

Around the World

- There are 1860 stairs. Therefore there could be:
  930 flights of two stairs
  620 flights of three stairs
  465 flights of four stairs
  372 flights of five stairs … etc.

AfL

- Is the hypothesis correct? Does the theorem work? Give me some examples to prove it.
- What is happening in this sequence?
- What pattern do you notice? What is the rule?
- How would you express this rule in words? What about using symbols?
- How else can you express this rule?
- What are all the different possibilities? How can you be sure that you have accounted for them all?

Answers

Page 1

Famous Mathematicians
The first twenty square numbers are: 1, 4, 9, 16, 25, 36, 49, 64, 81, 100, 121, 144, 169, 196, 225, 256, 289, 324, 361, 400.
Check children's answers to ensure that each of their chosen positive whole numbers can be written as the sum of four or fewer square numbers.

Let's Investigate
The hypothesis is correct.

Looking for Patterns
Add the previous two numbers and only write down the ones digit from the answer or +4, +2, –4, –2, +4, +2, …
Number sequences will vary.

Page 2

Looking for Patterns
Rule: Find the difference between the two numbers above.
The last number in each row is the difference between the first and last number above.
Whatever the numbers chosen, you always end up with four zeros.
If you start with:
two numbers – end with two zeros
three numbers – end with two ones and one zero.

Looking for Patterns
The numbers 1, 2, 3, 4, 6 and 8 end in the repeating pattern 1, 2, 1, 2, …
The numbers 5, 7 and 9 end in the repeating pattern 14, 7, 20, 10, 5, 14, 7, …
Number sequences will vary.

Technology Today
Explanations will vary.

Page 3

The Puzzler
Answers will vary.

Sports Update
301 ones are needed.
192 zeros are needed.
300 of each of the digits 2–9 are needed.
Therefore 2893 digits altogether are needed:
301 + 192 + (8 × 300).

Page 4

The Language of Maths
LCM: The smallest number that is exactly divisible by all of a given set of numbers.
It is useful to know the LCM when calculating with fractions.
Examples will vary.

The Puzzler

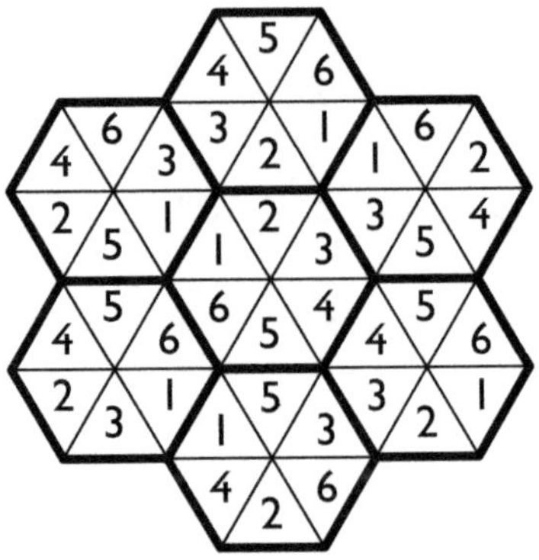

Other solutions are possible.

Around the World
There could be 2, 3, 4, 5, 6, 10, 12, 15, 20, 30, 31, 60, 62, 93, 124, 155, 186, 310, 372, 465, 620 or 930 stairs in each flight.
Results of the investigation will vary.

Inquisitive ant

numeral
A symbol or set of symbols used to represent a number, for example, the Arabic numeral 9, the equivalent Roman numeral IX, and the equivalent ancient Greek symbol ϑ.

Issue 2

Number

Prerequisites for learning

- Read Roman numerals to 1000 (M) and recognise years written in Roman numerals
- Identify prime numbers
- Understand square numbers, cube numbers and other powers
- Explore number sequences, patterns and relationships
- Use simple expressions and formulae in words and symbols

Resources

pencil and paper
Resource sheet 2: My notes (optional)
Resource sheet 3: Pupil self assessment booklet (optional)
calculator
computer with Internet access

Teaching support

Page 1

In the Past

- Ask the children to write some calculations using Braille. In order to do this they will need to know the following:

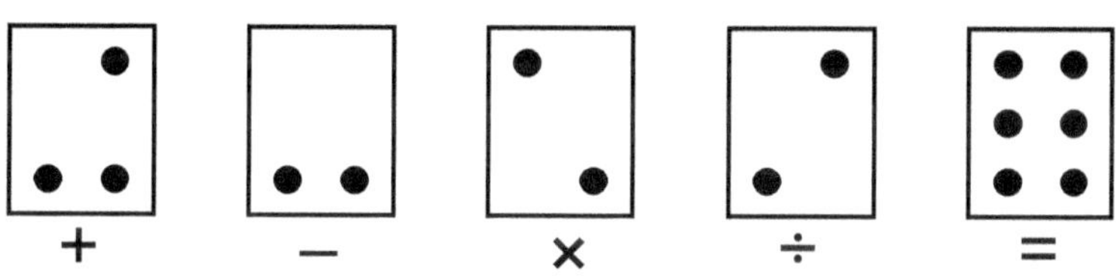

Page 2

Looking for Patterns

- The order of the relative frequencies of letters in the English language is generally agreed upon to be as follows: e, t, a, o, i, n, s, h, r, d, l, c, m, w, f, g, y, p, b, v, k, j, x, q and z.
- If necessary, tell the children that E followed by T are the two letters that appear most often in English language text.
- Children use Morse code to send a message to a friend.

What's the Problem?

- The largest Roman numeral you can write using each symbol once only is quite simple: it is the seven Roman numerals written in descending order.

Let's Investigate

- Children will probably need access to the Internet to find out the meaning of a bar (overline) above a Roman numeral. If this is not possible tell the children the Hindu-Arabic equivalent of $\overline{\text{V}}$ (i.e. 5000) and $\overline{\text{X}}$ (10 000). Can they identify a pattern in order to correctly write the Hindu-Arabic equivalent for the three other Roman numerals?
- Children write other numbers in Roman numerals using a bar (overline) above the numeral and give them to a friend to write the equivalent Hindu-Arabic numeral.

Page 3

Looking for Patterns

- The first two numbers in the sequences can be any two one-digit numbers. All the subsequent numbers are generated by adding the previous two numbers together and are therefore closely related to Fibonacci numbers.
- Suggest children find two given numbers in the sequence separated by one box. From this, they can work out what number goes in the box. They can then work backwards and forwards to find the other unknown numbers.

The Language of Maths

- Children who speak an additional language to English might investigate what happens in that language. Otherwise, children can investigate what happens for a language of either your or their choice.

Page 4

Let's Investigate

- Ensure that children correctly identify the primes to 100 before carrying out the investigations into twin primes.

Let's Investigate

- Is there a pattern to the tens digit, or the hundreds digit?

AfL

- What patterns / relationships do you notice? What conclusions can you draw?
- Can you give me a rule to describe this?
- Can you predict what the next number / the answer will be? What did you base your prediction on?

Answers

Page 1

In the Past
Braille message reads: 'I can read Braille'.
Messages and numbers will vary.

Page 2

Looking for Patterns
Answers will vary. In general however, the greater the frequency a letter is used in the English alphabet, the simpler the code for that letter. For example, the two letters that appear most often in English language text are E and T, and these have the simplest Morse identifiers: E (•) and T (—).
Although not all of the vowels have the simplest of Morse identifiers, A, E and I each only have one or two symbols, i.e. A (• —); E (•); I (• •).

What's the Problem?
I, V, X, L, C, D, and M
MDCLXVI (1666)
MCDXLIV (1444)

Let's Investigate

V	X	XXV	XLVI	LXXXIII
5	10	25	46	83

A conventional Roman numeral with a bar (overline) above the numeral is multiplied by 1000.

$\overline{V}$	$\overline{X}$	$\overline{XXV}$	$\overline{XLVI}$	$\overline{LXXXIII}$
5000	10 000	25 000	46 000	83 000

Page 3

Looking for Patterns
The next number in the sequence is the sum of the previous two numbers:
2, 5, 7, 12, 19, 31, 50, 81, 131
1, 6, 7, 13, 20, 33, 53, 86, 139
3, 9, 12, 21, 33, 54, 87, 141, 228

The Language of Maths
You always end up with four.
In French you end up with the repeating sequence 5, 4, 6, 3.
The repeating sequences vary for other languages.

Page 4

Let's Investigate
The prime numbers to 100 are: 2, 3, 5, 7, 11, 13, 17, 19, 23, 29, 31, 37, 41, 43, 47, 53, 59, 61, 67, 71, 73, 79, 83, 89, 97.

There are eight twin primes to 100: 3 & 5; 5 & 7, 11 & 13; 17 & 19; 29 & 31; 41 & 43; 59 & 61; 71 & 73.

3 + 5 = 8; 5 + 7 = 12; 11 + 13 = 24; 17 + 19 = 36;
29 + 31 = 60; 41 + 43 = 84; 59 + 61 = 120; 71 + 73 = 144.
With the exception of the first twin primes (3 + 5 = 8), the sums of all the other twin primes are divisible by 12.

3 × 5 = 15; 5 × 7 = 35, 11 × 13 = 143; 17 × 19 = 323;
29 × 31 = 899; 41 × 43 = 1763; 59 × 61 = 3599;
71 × 73 = 5183.
The product of each twin prime is one less than the square of the number that lies between the two numbers of the twin prime, i.e.
3 × 5 = 15
$4^2 = 16$
16 – 1 = 15

Let's Investigate
For 3 =

For 4 = 4 ➜ 6
For 5 = 5
For 6 = 6
For 7 =
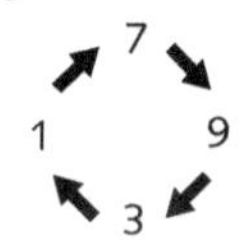

For 8 =
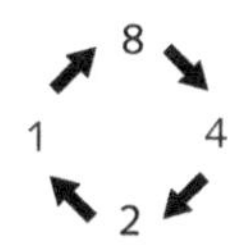

For 9 = 9 ➜ 1
For 10 = 0

Inquisitive ant

conjecture
A conclusion, judgement, or statement based on inconclusive or incomplete information.

Issue 3

Number

Prerequisites for learning

- Explore number sequences, patterns and relationships
- Read Roman numerals to 1000 (M) and recognise years written in Roman numerals
- Identify prime numbers
- Recognise and use proper factors
- Understand and use tests of divisibility
- Recognise and use square numbers

Resources

pencil and paper
Resource sheet 2: My notes (optional)
Resource sheet 3: Pupil self assessment booklet (optional)
Resource sheet 19: 2 cm squared paper
range of different theatre plans – available on the Internet (optional)
computer with Internet access

Teaching support

Page 1

The Puzzler

- Children investigate consecutive numbers with two proper factors (14 and 15; 21 and 22; 26 and 27; 33, 34 and 35; 38 and 39; 57 and 58; 85, 86 and 87; 93, 94 and 95).

Let's Investigate

- Children investigate which prime numbers to 1000 are also prime numbers when their digits are reversed (101, 107, 113, 131, 149, 151, 157, 167, 179, 181, 191, 199, 311, 313, 337, 347, 353, 359, 373, 383, 389, 701, 709, 727, 733, 739, 743, 751, 761, 769, 787, 797, 907, 919, 929, 937, 941, 953, 967, 971, 983, 991)

Let's Investigate

- Children investigate pairs of consecutive odd numbers between 200 and 400 that are both primes (227 and 229; 239 and 241; 269 and 271; 281 and 283; 311 and 313; 347 and 349).

Page 2

Looking for Patterns

- Assist the children in seeing the following patterns:
  - the ones and thousands digits are even numbers
  - the sum of all four digits is divisible by 3, e.g.
    2112 = 2 + 1 + 1 + 2 = 6; 4554 = 4 + 5 + 5 + 4 = 18

Looking for Patterns

- Children investigate which two-digit numbers have exactly twelve factors (48 and 80).

The Puzzler

- Tell the children that there are three different combinations for making four square numbers (one one-digit number, one two-digit number and two three-digit numbers) and only one combination for making three square numbers (one two-digit number, one three-digit number and one four-digit number).

Page 3

The Arts Roundup

- Discuss the activity with the children and, if appropriate, have available a range of different theatre plans. Children need to design the auditoriums with exactly the number of seats given. Encourage them to think about the factors of 840 and 1260, and to ensure that the seating in the auditoriums is 'symmetrical'.

What's the Problem?

- The key to identifying the correct number of factors is based on factor pairs and whether or not the number is a square number. If it is not a square number, then there should be an even number of factors. If it is a square number, then its square root is paired with itself and there are an odd number of factors.

Page 4

The Puzzler

- Tell the children that there are six possible three-digit numbers for each puzzle. Can they find them all?

In the Past

- There are a range of different number systems that the children could investigate such as: Babylonians, Ancient Egyptian, Ancient Greek, Ancient Chinese, Ancient Japanese and Hebrew.

AfL

- What are all the different possibilities? How can you be sure that you have accounted for them all?
- What is happening in this sequence?
- What pattern do you notice? What is the rule?
- How would you express this rule in words? What about using symbols?
- How else can you express this rule?

Answers

Page 1

The Puzzler
44 and 45
98 and 99

Let's Investigate
11, 13, 17, 31, 37, 71, 73, 79, 97

Let's Investigate
101 and 103
107 and 109
137 and 139
149 and 151
179 and 181
191 and 193
197 and 199

Page 2

Looking for Patterns
In each number:

- the ones and thousands digits are the same
- the tens and hundreds digits are the same.

There are ten other numbers that are divisible by 66 and have the same ones / thousands and tens / hundreds patterns: 2442, 2772, 4224, 4884, 6006, 6336, 6996, 8118, 8448, 8778.

Looking for Patterns
60, 72, 90 and 96

The Puzzler
There are three different combinations for making four square numbers:
9, 81, 324, 576
9, 25, 361, 784
1, 36, 529, 784
There is only one combination for making three square numbers: 25, 841, 7396

Page 3

The Arts Roundup
Designs will vary.

What's the Problem?
42 is missing.
If 1008 (or any number) is not a square number, then there should be an even number of factors. If the number is a square number, then its square root is paired with itself and there is an odd number of factors.
As 1008 is not a square number there should be an even number of factors. By pairing up the factors, e.g. 1 × 1008, 2 × 504 … it can be seen that one of the factors does not have a pair, i.e. 24. Therefore 42 is missing.

Page 4

The Puzzler
123, 132, 213, 231, 312, 321
257, 275, 527, 572, 725, 752

In the Past
Results of the investigation will vary.

Inquisitive ant

natural number
A whole number greater than zero.

Issue 4

Negative numbers

Prerequisites for learning

- Relate numbers to their position on a number line
- Interpret negative numbers in context, count forwards and backwards with positive and negative whole numbers, including through zero
- Use a number line to add and subtract positive and negative integers in context, including for measures such as temperature
- Construct and interpret frequency tables, bar charts and line graphs
- Calculate mentally with integers

Resources

pencil and paper
Resource sheet 2: My notes (optional)
Resource sheet 3: Pupil self assessment booklet (optional)
ruler
coloured pencils
calculator (optional)
computer

Teaching support

Page 1

Around the World

- If children have difficulty with this activity, mark temperatures from pairs of cities on the same thermometer and let them count the differences.

The Puzzler

- Ensure that children understand that subtracting a negative number means adding that number, e.g. $4 - {}^{-}2 = 6$.

Page 2

Looking for Patterns

- Work with the children to fill in some of the boxes in the table or complete a row and a column (see Answers). Discuss with them the rules for multiplying different combinations of positive and negative integers.
- If the children complete the grid correctly, when they colour it they should be able to draw conclusions and identify the rules more easily.

Let's Investigate

- Suggest the children use a number line to help them identify pairs of numbers with a difference of four.

Page 3

The Puzzler

- Ensure children are familiar with magic squares.
- Point out that it is much easier to make up a magic square if one of the corner numbers is zero.

Around the World

- Ensure the children understand the difference between the 'average maximum' and 'absolute maximum', and 'average minimum' and 'absolute minimum'.

Page 4

Around the World

- Ensure the children understand the terms 'average daily temperature (max)' and 'average daily temperature (min)'.

Let's Investigate

- Suggest the children use a number line to help them add sets of positive and negative integers totalling zero.

AfL

- What is the difference between these two numbers?
- What is the rule for finding the difference between a positive and a negative number? What about the difference between two negative numbers?
- What about adding together a positive number and a negative number?
- How does a number line help when adding and subtracting positive and negative numbers?
- What rules can you tell me about multiplying different combinations of positive and negative numbers? Give me some examples. What patterns did you notice? How did your knowledge of the times-tables help?
- What does this graph tell us? Is this the best way to show this? How else could you have displayed the information?

Answers

Page 1

Around the World
Statements will vary.

The Puzzler

7	4	-2	5	3
3	6	-7	2	
-3	13	-9		
-16	22			
-38				

Page 2

Looking for Patterns

×	-5	-4	-3	-2	-1	0	1	2	3	4	5
5	-25	-20	-15	-10	-5	0	5	10	15	20	25
4	-20	-16	-12	-8	-4	0	4	8	12	16	20
3	-15	-12	-9	-6	-3	0	3	6	9	12	15
2	-10	-8	-6	-4	-2	0	2	4	6	8	10
1	-5	-4	-3	-2	-1	0	1	2	3	4	5
0	0	0	0	0	0	0	0	0	0	0	0
-1	5	4	3	2	1	0	-1	-2	-3	-4	-5
-2	10	8	6	4	2	0	-2	-4	-6	-8	-10
-3	15	12	9	6	3	0	-3	-6	-9	-12	-15
-4	20	16	12	8	4	0	-4	-8	-12	-16	-20
-5	25	20	15	10	5	0	-5	-10	-15	-20	-25

Two positive numbers multiplied together make a positive.
Two negative numbers multiplied together make a positive.
A positive and a negative number multiplied together make a negative.

Let's Investigate
-3 & 1, -2 & 2, -1 & 3
-1 & -5, -2 & -6, -3 & -7 … etc.
Difference of 6: -5 & 1 to -1 & 5
Difference of 10: -9 & 1 to -1 & 9
Difference of 15: -14 & 1 to -1 & 14.

Page 3

The Puzzler

-7	4	-6
-2	-3	-4
0	-10	1

Magic number is -9.
Other solutions are possible.

Around the World
Reports will vary.

Page 4

Around the World
Reports will vary.

Let's Investigate
Results of the investigation will vary.

Inquisitive ant

integer
Any positive or negative whole number or zero.

Issue 5

Addition

Prerequisites for learning

- Explore number sequences, patterns and relationships
- Use simple expressions and formulae in words and symbols
- Calculate mentally with integers and decimals
- Use the formal written method to add whole numbers and decimals

Resources

pencil and paper
Resource sheet 2: My notes (optional)
Resource sheet 3: Pupil self assessment booklet (optional)
counters (optional)
calculator
set of dominoes
pack of playing cards

Teaching support

Page 1

The Puzzler

- As it is not possible to make an odd total, there must always be only two odd numbers face up – never one or three. This means that 9 and 7 must have odd numbers on the back, and 6 an even number.
- The highest number possible is 22 (9 + 6 + 7), therefore the unknown numbers must all be smaller than the numbers we know. So, the number on the back of 6 must be 2 or 4.
- The smallest total is 10, therefore:
  - if 2, the other numbers must be 3 and 5 (not 7 and 1 as there is already a 7)
  - if 4, the other numbers must be 5 and 1 (not 3 and 3).
- Encourage the children to narrow down the possible numbers first, using the information that is given. If necessary, give them three counters with 9, 6 and 7 written on them and ask them to write odd or even on the back of each of them and then toss them to see what combination always comes out even.

Let's Investigate

- Ask the children to comment on any patterns they notice in the numbers.

Money Matters

- What is the least number of coins needed to be able to make any amount of money from 1p to £1? (1p, 2p, 2p, 5p, 10p, 10p, 20p, 50p).

The Puzzler

- Ensure children are familiar with magic squares.
- Encourage the children to work in pairs to create their own magic squares.

Page 2

In the Past

- Ensure children are familiar with Roman numerals and magic squares.
- Encourage the children to work in pairs to create their own magic squares.

The Puzzler

- Ensure the children realise that each side of the domino constitutes a number in the magic square. Therefore only eight dominoes are needed to make each magic square.
- There are many different 4 × 4 magic squares that can be created using a set of dominoes.

Looking for Patterns

- Work with the children to start off the investigation. Once they have identified the pattern, leave them to continue independently to write and apply the formula.
- Once they have completed this activity, children should be encouraged to start on the next Looking for Patterns activity on page 3.

Page 3

Looking for Patterns

- Ensure that the children have completed and fully understood the Looking for Patterns activity on page 2 before starting on this activity.
- Work with the children to start off the investigation. Once they have identified the pattern, leave them to continue independently to write and apply the formula.

Money Matters

- Once the children have completed this activity, discuss with them the strategies they used to work out which credit card receipts went with which bills.

The Language of Maths

- This activity requires children to find the number of pages with a 2 in the page number, not the total number of 2s in the page numbers, e.g. page 222 counts as one.
- From pages 1 to 100 the number 2 can be found on the following pages:
  2, 12, 20, 21, 22, 23, 24, 25, 26, 27, 28, 29, 32, 42, 52, 62, 72, 82, 92 (19 pages)
  So:

Pages 1–100: 19 pages	Pages 501–600: 19 pages
Pages 101–199: 19 pages	Pages 601–700: 19 pages
Pages 200–299: 100 pages	Pages 701–800: 19 pages
Pages 300–400: 19 pages	Pages 801–900: 19 pages
Pages 401–500: 19 pages	Pages 901–1000: 19 pages

  Pages 1001–1081: 17 pages (1002, 1012, 1020, 1021, 1022, 1023, 1024, 1025, 1026, 1027, 1028, 1029, 1032, 1042, 1052, 1062, 1072)
  Total: 288

Page 4

The Puzzler

- Children investigate what answers they can make using any of the four operations and brackets. What's the smallest / largest answer they can make? What about an answer as close to 100 as possible?

Let's Investigate

- Ensure children understand how an addition tower (or number wall) functions.
- Ask children to find the rule for the relationship between the middle number in the bottom row and the top number of any addition tower with an odd number of numbers in the bottom row.
  (If there are x numbers in the bottom row of an addition tower and x is odd, then the top number (T) will be $(x - 1)^2$ times bigger than the middle number (n) in the bottom row.
  $T = n(x - 1)^2$, if x is odd.)

The Puzzler

- What about a square with eight numbers or a pentagon with ten numbers?

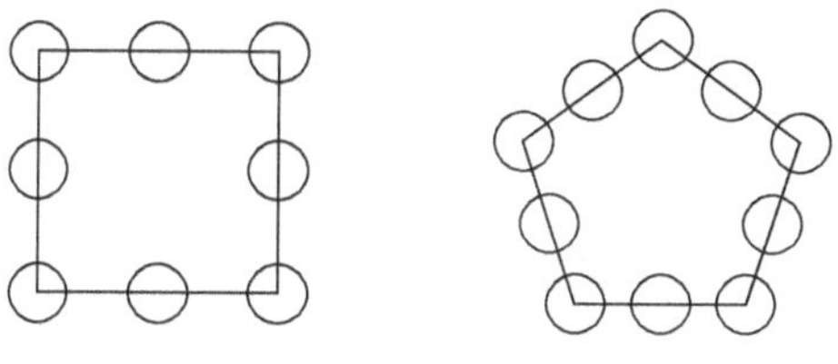

AfL

- How did you work out the solution to this puzzle? What strategies did you use? How else could you have gone about it?
- Did you use trial and improvement to help you work out the answer to this puzzle? How did you know where to start? Then what did you do?
- How did you organise your thoughts? Did this help? What else could you have done? Why might this have been more effective?
- Have you accounted for all the different possibilities? How can you be sure?
- What patterns did you notice? What rules can you make?

Answers

Page 1

The Puzzler
The other numbers written on the counters are 5 on the back of 9, 4 on the back of 6 and 1 on the back of 7.
Or 3 on the back of 9, 2 on the back of 6 and 5 on the back of 7.

Let's Investigate
105, 114, 123, 132, 141, 150
204, 213, 222, 231, 240
303, 312, 321, 330
402, 411, 420
501, 510
600

Money Matters
1p, 2p, 3p, 5p, 6p, 7p, 8p,
10p, 11p, 12p, 13p, 15p, 16p, 17p, 18p,
20p, 21p, 22p, 23p, 25p, 26p, 27p, 28p,
30p, 31p, 32p, 33p, 35p, 36p, 37p, 38p,
50p, 51p, 52p, 53p, 55p, 56p, 57p, 58p,
60p, 61p, 62p, 63p, 65p, 66p, 67p, 68p,
70p, 71p, 72p, 73p, 75p, 76p, 77p, 78p,
80p, 81p, 82p, 83p, 85p, 86p, 87p, 88p,
£1, £1.01, £1.02, £1.03. £1.05, £1.06, £1.07, £1.08,
£1.10, £1.11, £1.12, £1.13, £1.15, £1.16, £1.17, £1.18,
£1.20, £1.21, £1.22, £1.23, £1.25, £1.26, £1.27, £1.28,
£1.30, £1.31, £1.32, £1.33, £1.35, £1.36, £1.37, £1.38,
£1.50, £1.51, £1.52, £1.53, £1.55, £1.56, £1.57, £1.58,
£1.60, £1.61, £1.62, £1.63, £1.65, £1.66, £1.67, £1.68,
£1.70, £1.71, £1.72, £1.73, £1.75, £1.76, £1.77, £1.78,
£1.80, £1.81, £1.82, £1.83, £1.85, £1.86, £1.87, £1.88
The amounts not listed above cannot be made using the coins given.

The Puzzler
Magic squares will vary.

Page 2

In the Past

The magic number is XXXIII

VIII	XVI	IX
XII	XI	X
XIII	VI	XIV

The magic number is XXIV

XI	IV	IX
VI	VIII	X
VII	XII	V

The Puzzler

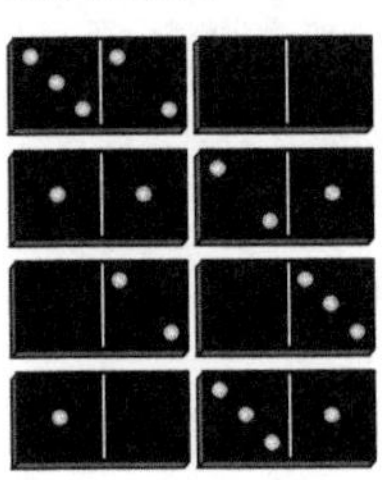

Many other solutions are possible.

Looking for Patterns
Total of the first two odd numbers: 4 (1 + 3)
Total of the first three odd numbers: 9 (1 + 3 + 5)
Total of the first four odd numbers: 16 (1 + 3 + 5 + 7)
Total of the first five odd numbers: 25 (1 + 3 + 5 + 7 + 9)
4, 9, 16 and 25 are the answers to the squares from 2 × 2 to 5 × 5.
The total (T) of the first n consecutive odd numbers is equal to the square of the number of odd numbers, $T = n^2$.
Therefore, the total of the first 100 odd numbers is $100^2 = 10\ 000$.

Page 3

Looking for Patterns
Total of the first two even numbers: 6 (2 + 4)
Total of the first three even numbers: 12 (2 + 4 + 6)
Total of the first four even numbers: 20 (2 + 4 + 6 + 8)
Total of the first five even numbers: 30 (2 + 4 + 6 + 8 + 10)
6 = 2 even numbers × 3
12 = 3 even numbers × 4
20 = 4 even numbers × 5
30 = 5 even numbers × 6
The total (T) of the first n consecutive even numbers is equal to the number of even numbers multiplied by the number of even numbers plus one, $T = n \times (n + 1)$.
Therefore, the total of the first 100 even numbers is 100 × 101 = 10 100.

Money Matters
Total £83.42 = £46.18 + £37.24
Total £168.60 = £56.20 + £56.20 + £56.20
Total £75.33 = £36.50 + £38.83
Total £204.86 = £70 + £66.50 + £68.36

The Language of Maths
288

Page 4

The Puzzler
13 + 12 − 9 + 7 = 23
13 + 12 + 9 + 7 = 41
13 + 12 + 9 − 7 = 27
13 − 12 + 9 − 7 = 3
13 − 12 + 9 + 7 = 17

Let's Investigate
The statement is true. $T = 4n$

The Puzzler

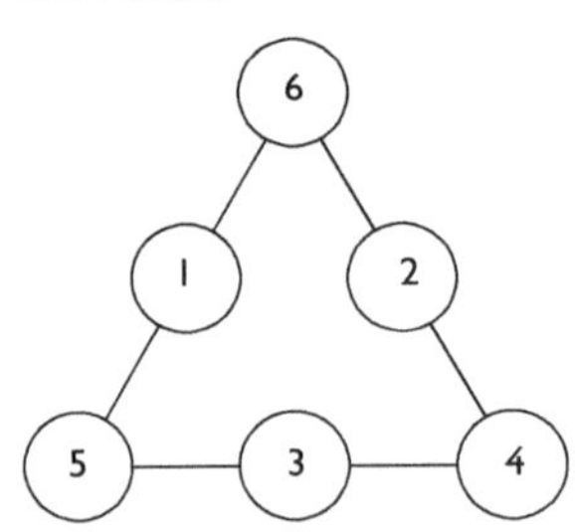

The numbers in the corners are either the highest or the lowest consecutive numbers.

0–5 and consecutive even numbers will work, as will any sequence of six numbers with the same difference between each one.

Other solutions are possible.

Inquisitive ant

∞
The sign for infinity.
Infinity refers to an amount or number that is so great that it cannot be counted.

Issue 6

Subtraction

Prerequisites for learning

- Recognise and use square numbers
- Identify prime numbers
- Read, write, order and compare numbers with up to three decimal places
- Understand and use the order of operations
- Begin to use brackets and know that they determine the order of operations
- Calculate mentally with integers and decimals
- Use the formal written method to subtract whole numbers and decimals

Resources

pencil and paper
Resource sheet 2: My notes (optional)
Resource sheet 3: Pupil self assessment booklet (optional)
calculator
material for making a poster
lists of three-digit prime and square numbers (optional)
computer

Teaching support

Page 1

What's the Problem?

- Once children have answered the question, ask them to find out what Malcolm's mileometer will read the next time the digits 1, 3, 4 and 6 appear (1643), and how many more miles Malcolm will have travelled? (9 miles). What about the next time the digits appear and the time after that …?

Let's Investigate

- Ensure that the children are familiar with the terms 'minuend', 'subtrahend' and 'difference'. If appropriate, ask the children to complete The Language of Maths activity on page 4 prior to this activity.

Page 2

The Puzzler

- Ensure that the children realise that brackets can go around three of the numbers in the calculation.

Let's Investigate

- To get the smallest difference between the largest and smallest number, the smallest number must be of the form 0·*x*98 and the largest number of the form 0·*z*12, with *y* being the tenths digit of the middle number and *x*, *y* and *z* being the three consecutive digits between 3 and 7. So the smallest and largest numbers must be:
  0·398 and 0·512, or
  0·498 and 0·612, or
  0·598 and 0·712.
- To get the largest difference, the smallest number must be 0·123 and the largest 0·987: the difference being 0·864.

Let's Investigate

- Ensure children understand that the number in the middle of each line is the difference between the two numbers at either end of the line.

Page 3

Let's Investigate

- This activity requires children to use the problem solving strategy of accounting for all possibilities. It is important therefore that children are systematic in their recording.

Looking for Patterns

- Children should have few problems in identifying and continuing the pattern that Abigail used to write the set of calculations. Nor should they have difficulties in calculating the differences between pairs of answers. However, with regards to commenting on any patterns they notice, it is only expected that children comment on the recurring pattern of the digits 0, 1 and 9.

Let's Investigate

- Similar to the Let's Investigate activity at the top of page 3, this activity requires children to account for all the possibilities. However, children are then expected to identify pairs of calculations with the same total.

Page 4

The Puzzler

- One strategy is to divide the total of each addition calculation by the number of boxes that add together to make it: this gives the middle number in the sequence. In the case of the second calculation, which has six boxes, this gives the number half way between the two centre numbers i.e. 17·5, therefore the numbers on either side are 16 and 19 (a difference of 3 between them). Once the middle number(s) have been identified, the numbers on either side can then easily be worked out by adding or subtracting two, three or four.

The Language of Maths

- Explain to the children that their poster will go on display in the classroom and that it should be designed so that other children can easily understand the meaning of each of the words.

The Puzzler

- Provide the children with lists of three-digit prime and square numbers.

AfL

- What is the answer to the problem? How did you work it out? What strategies did you use?
- What is the solution to this puzzle?
- Use your own words to explain to me this method of subtraction.
- Is this a method that you might use regularly? Why? / Why not?
- How would you work out the answer to this calculation? Why would you use this method?
- Have you accounted for all the possibilities? How do you know?

Answers

Page 1

What's the Problem?

The next time the digits 1, 3, 4 and 6 appear on Malcolm's mileometer it will read: 1634.
Therefore, Malcolm has to travel another 171 miles.

Let's Investigate

Results of the investigation will vary.
This method works for four-digit numbers if you change the minuend to 9999.

Page 2

The Puzzler

(832 – 65) – 142 + 152 = 777 or (832 – 65 – 142) + 152 = 777
832 – (65 – 142) + 152 = 1061
832 – 65 – (142 + 152) = 473
832 – (65 – 142 + 152) = 757

Let's Investigate

The difference between Sharon's smallest and largest numbers is 0·684.

The smallest difference possible between the smallest and largest numbers is 0·114.
The largest difference is 0·864.

Let's Investigate

The next stage of the diagram has four zeros.
The innermost square always has four zeros but you may need to draw several more squares to get there.

Page 3

Let's Investigate

9·5 – 8·76 or 8·67, 9·6 – 8·75
8·5 – 7·96 or 7·69, 8·6 – 7·95
7·5 – 6·98 or 6·89
6·7 – 5·98 or 5·89
6·8 – 5·97

Looking for Patterns

12 345 – 6789 = 5556
23 456 – 7891 = 15 565
34 567 – 8912 = 25 655
45 678 – 9123 = 36 555
56 789 – 1234 = 55 555
The next two calculations are:
67 891 – 2345 = 65 546
78 912 – 3456 = 75 456
Differences between pairs of calculations are:
15 565 – 5556 = 10 009
25 655 – 15 565 = 10 090
36 555 – 25 655 = 10 900
55 555 – 36 555 = 19 000
65 546 – 55 555 = 9991
75 456 – 65 546 = 9910

Let's Investigate

9·8 – 5·67 = 4·13 and 9·8 – 7·65 = 2·15
9·8 – 5·76 = 4·04 and 9·8 – 7·56 = 2·24
9·8 – 6·57 = 3·23 and 9·8 – 6·75 = 3·05
The sum of each pair of totals is 6·28.

9·87 – 5·6 = 4·27 and 9·85 – 7·6 = 2·25
9·86 – 5·7 = 4·16 and 9·86 – 7·5 = 2·36
9·87 – 6·5 = 3·37 and 9·85 – 6·7 = 3·15
The sum of each pair of totals is 6·52.

Page 4

The Puzzler

16 + 18 + 20 + 22 + 24 = 100
10 + 13 + 16 + 19 + 22 + 25 = 105
16 + 20 + 24 + 28 + 32 = 120

The Language of Maths

minuend
– subtrahend
difference

The Puzzler

307	prime	557	prime	576	square
– 144	– square	– 196	– square	– 269	– prime
163	prime	361	square	307	prime

Other answers are possible.

Inquisitive ant

differential
Can mean either a difference between two values on a scale, for example, a difference in exchange rates; or an extremely small difference.

Issue 7

Multiplication

Prerequisites for learning

- Identify pairs of factors of two-digit whole numbers and find the prime factors of two-digit numbers
- Understand square numbers, cube numbers and other powers
- Understand the relationships between operations, including inverse operations, and how these can be used to calculate more efficiently
- Use the formal written methods of short multiplication and long multiplication
- Use approximations, inverse operations and tests of divisibility to estimate and check results

Resources

pencil and paper

Resource sheet 2: My notes (optional)

Resource sheet 3: Pupil self assessment booklet (optional)

Resource sheet 4: 9 × 9 grids (preferably enlarged to A3)

calculator

stopwatch

Teaching support

Page 1

In the Past

- Encourage the children to use the Russian peasant method for multiplying pairs of three-digit or four-digit numbers. What about using decimal numbers?

Let's Investigate

- Encourage the children to choose three consecutive three-digit numbers, then four-digit numbers.

Page 2

In the Past

- Ensure children understand the meaning of the term 'digital root' and are able to find the digital root for any two-digit number.
- Ask the children to join together all of the squares with a 5, 6, 7, 8 or 9 digit in them.

The Puzzler

- Ensure children understand the meaning of the terms 'digital root' and 'cubed numbers', and can calculate cubed numbers.

Page 3

The Puzzler

- Multiplying by 7, 11 and 13 is the same as multiplying by 1001 (i.e. 7 × 11 × 13 = 1001). Multiplying by 1000 moves all the digits in the number three places to the left (168 000). Multiplying by 1 results in the same digits being repeated in the hundreds, tens and ones place values (168 000 + 168).

Sports Update

- Children should measure their heart rates for a minute using stopwatches. They then scale up to work out the approximate number of times it beats in a year.
- The average heart rate for an adult is around 70 beats per minute.

Looking for Patterns

- Ensure children understand the meaning of the term 'cubed numbers', and can calculate both square and cubed numbers.

The Puzzler

- Using a calculator or written method of division: Try dividing 17 280 by the list of given numbers, starting with the largest. Then use knowledge of tests of divisibility to find other factors from the given numbers.
- Without a calculator: find the prime factors by dividing, i.e. $2^7 \times 3^3 \times 5$, and then combine these to make the numbers on the given list.

Page 4

Looking for Patterns

- From the second calculation it is clear that S = 0, E × Y = 8, so E and Y must be 2 and 4 (not 1 and 8, as we are given 8).
  W × E = □E, so W must be 1 or 6.
  W × 8 + □ = TS so W must be 6, T must be 5 and □ = 2, which means that E = 4 and therefore Y = 2.
  Put these numbers into the first calculation and the other letters in both calculations can then be found.
- Tell children what one or more of the digits represents, e.g. Y = 2.

The Arts Roundup

- There are 23 rows in the theatre.

Looking for Patterns

- A simple way to start to estimate the magnitude of the numbers is by referring to the number of zeros in each calculation.

AfL

- Explain to me this method for multiplication. What do you think of this method? Is it a method that you might ever use? Why? / Why not?
- What do you notice? What is the pattern? What generalisations can you make?
- What answer did you get for this puzzle? When you first looked at the puzzle how did you begin to work it out?
- Talk me through your method of working out the puzzle / problem.
- How did you order these numbers / calculations without first working out the answers? What things did you look for?

Answers

Page 1

In the Past
Results of the investigation will vary.

Let's Investigate
The square of the middle number is always one more than the product of the other two numbers.

Page 2

In the Past

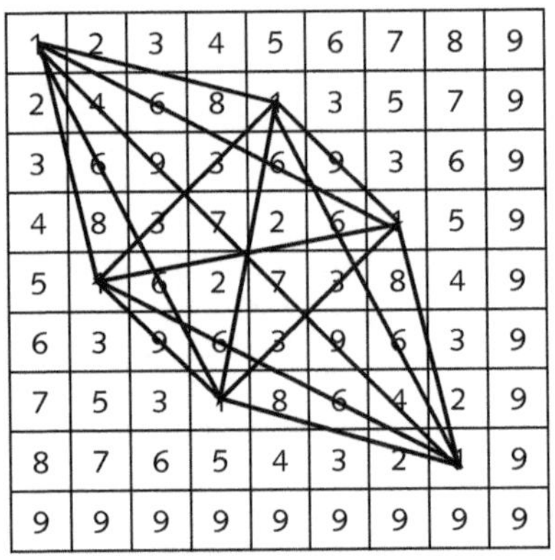

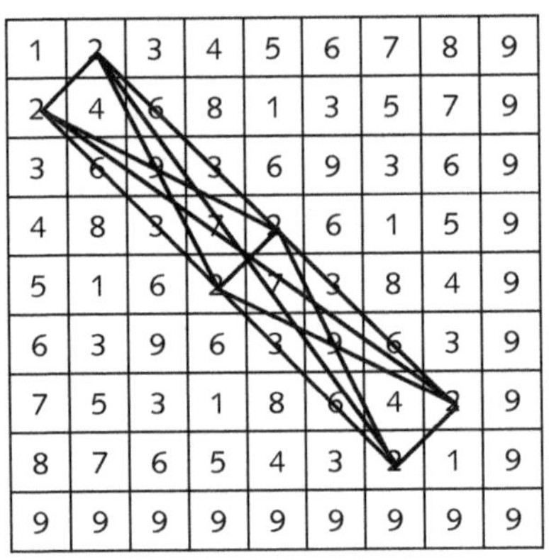

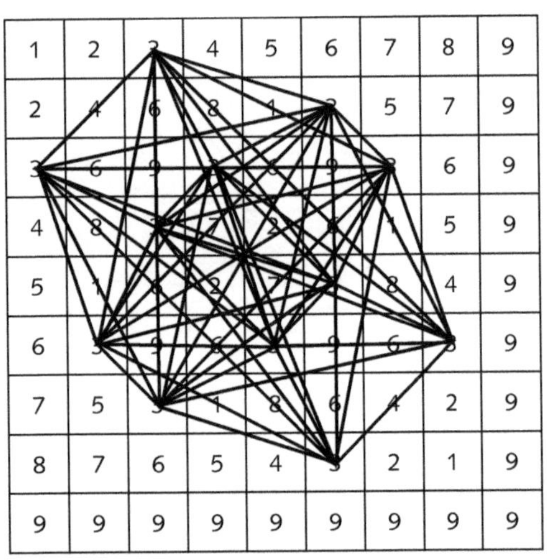

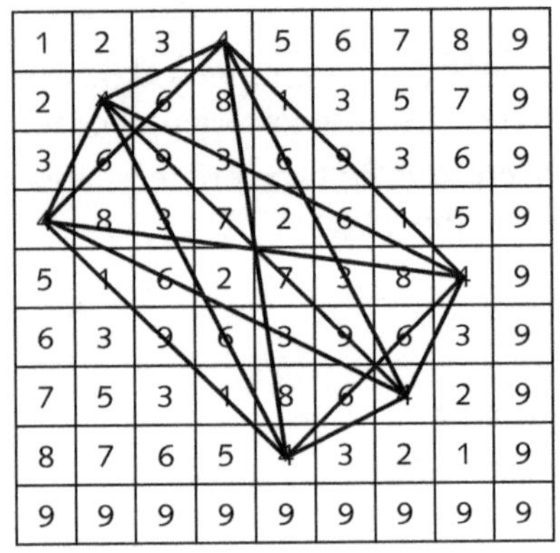

The Puzzler
The digits 1, 8 and 9 repeat in sequence.

Page 3

The Puzzler
The final answer will always be the three digits in a three-digit number written twice, e.g. 168 168.

Sports Update
Results of the investigation will vary.
The average adult's heart beats 36 792 000 times in a year.

Looking for Patterns
$2 \times 3 \times 4 \times 5 + 1 = 121 = 11^2$
$3 \times 4 \times 5 \times 6 + 1 = 361 = 19^2$
$4 \times 5 \times 6 \times 7 + 1 = 841 = 29^2$
$5 \times 6 \times 7 \times 8 + 1 = 1681 = 41^2$
$6 \times 7 \times 8 \times 9 + 1 = 3025 = 55^2$
The final number is the product of the middle two multiples minus 1, e.g. $11 = 3 \times 4 - 1$.

$1^3 + 2^3 + 3^3 = 36 = 6^2$
$1^3 + 2^3 + 3^3 + 4^3 = 100 = 10^2$
$1^3 + 2^3 + 3^3 + 4^3 + 5^3 = 225 = 15^2$
$1^3 + 2^3 + 3^3 + 4^3 + 5^3 + 6^3 = 441 = 21^2$
$1^3 + 2^3 + 3^3 + 4^3 + 5^3 + 6^3 + 7^3 = 784 = 28^2$
The final number is the sum of all the numbers added together, before they are cubed, e.g. $1 + 2 + 3 = 6$.

The Puzzler
$4 \times 5 \times 8 \times 9 \times 12$ or $2 \times 3 \times 4 \times 6 \times 10 \times 12$.

Page 4

Looking for Patterns
$3564 \times 7 = 24\,948$

$$\begin{array}{rl} 84 & \\ \times\ 62 & \\ \hline 168 & (84 \times 2) \\ 5040 & (84 \times 60) \\ \hline 5208 & \end{array}$$

The Arts Roundup
There are 391 seats in the theatre.

Looking for Patterns
$2^6 \times 10 = 640$
$0{\cdot}2 \times 6000 = 1200$ / $2000 \times 0{\cdot}6 = 1200$
$2^6 \times 6^2 = 2304$
$20^3 = 8000$
$200 \times 600 = 120\,000$
$2 \times 200 \times 2000 = 800\,000$
$10^6 = 1\,000\,000$
$60\,000 \times 200 = 12\,000\,000$
$6^{10} = 60\,466\,176$

Inquisitive ant

William Oughtred
An English mathematician (1574–1660) who assigned the symbol '×' to the multiplication operation and invented the slide rule.

Issue 8

Division

Prerequisites for learning

- Recognise and use square numbers
- Identify prime numbers
- Identify triangular numbers
- Understand tests of divisibility
- Identify pairs of factors of two-digit whole numbers and find common multiples
- Read, write, order and compare numbers with up to three decimal places
- Understand the relationships between operations, including inverse operations, and how these can be used to calculate more efficiently
- Divide numbers up to four digits by a one-digit number using the formal written method of short division
- Begin to divide numbers up to four digits by a two-digit number using the formal written method of long division
- Use approximations, inverse operations and tests of divisibility to estimate and check results

Resources

pencil and paper
Resource sheet 2: My notes (optional)
Resource sheet 3: Pupil self assessment booklet (optional)
two, three and four digit cards (optional)
calculator
material for making a poster
lists of one-, two- and three-digit primes, squares and triangular numbers (optional)
computer

Teaching support

Page 1

Money Matters

- The amount of money in the piles of four £1 coins is £1 less than in the piles of five £1 coins. Therefore children need to find a multiple of four that is one less than a multiple of five, i.e. a multiple of four that ends in four (there are no multiples of four that end in nine) and then add on £2.

What's the Problem?

- Discuss with the children how they worked out the answer. Did they find the lowest common multiple of 8 and 12, i.e. 24, and use this to work out the different combinations?
- Ask the children to find a rule for working out the number of tables needed for any given combination of table seatings, e.g. tables that seat six or ten children.

What's the Problem?

- Suggest the children draw a table to work out all the possible combinations.

Page 2

The Puzzler

- Tell the children that each of the three calculations involves TO·t divided by O.
- Tell the children each of the divisors (see Answers).

What's the Problem?

- Suggest the children use several of each of the digit cards 2 (to represent bicycles), 3 (to represent tricycles) and 4 (to represent cars), e.g. [4] [4] [4] [4] [4] [2].

Let's Investigate

- Ensure children are familiar with, and are able to draw, factor trees.

The Language of Maths

- Ensure children know the meaning of each of the words.

Page 3

Looking for Patterns

- Work through some numbers with the children to get them started, i.e.

$1^2 \div 3$	$2^2 \div 3$	$3^2 \div 3$	$4^2 \div 3$
$= 0$ r 1	$= 4 \div 3$	$= 9 \div 3$	$= 16 \div 3$
	$= 1$ r 1	$= 3$	$= 5$ r 1

Let's Investigate

- Tell the children each of the divisors (see Answers).

Let's Investigate

- Remind the children that:
  - a number is divisible by three if the sum of all the digits is divisible by three
  - a number is divisible by six if the number is even and divisible by three or if the number is divisible by both two and three
  - a number is divisible by eight if the last three digits are divisible by eight.

Looking for Patterns

- Remind the children that a number is divisible by nine if the sum of all the digits is divisible by nine.

Page 4

Looking for Patterns

- Discuss with the children an alternative method for determining whether a number is divisible by 11, i.e. subtract the last digit from the rest of the number. Continue to do this until you come to a number you know is, or is not, divisible by 11. For example:

  6457

  645 – 7 = 638

  63 – 8 = 55

  Since 55 is divisible by 11, then so also is 6457.

The Puzzler

- Provide children with lists of one-digit, two-digit and three-digit primes, squares and triangular numbers.

AfL

- Did you find all the possible combinations? Why are you so sure?
- What patterns did you notice?
- How did you work out the answer? Did you make an approximation? What did you do? How did your approximations help you find the right answer?
- Did you use trial and improvement to help you work out the answer to this puzzle? Where did you begin? Why did you start there?
- How did knowing about multiples / factors / divisors help you with this problem / puzzle? What about applying tests of divisibility?
- What patterns did you look for when going about answering this problem? How did these help?

Answers

Page 1

Money Matters
£6 or £26.
Other answers are possible.

What's the Problem?
1 table of 8 and 36 tables of 12
4 tables of 8 and 34 tables of 12
7 tables of 8 and 32 tables of 12
10 tables of 8 and 30 tables of 12
13 tables of 8 and 28 tables of 12
16 tables of 8 and 26 tables of 12
19 tables of 8 and 24 tables of 12
22 tables of 8 and 22 tables of 12
25 tables of 8 and 20 tables of 12
28 tables of 8 and 18 tables of 12
31 tables of 8 and 16 tables of 12
34 tables of 8 and 14 tables of 12
37 tables of 8 and 12 tables of 12
40 tables of 8 and 10 tables of 12
43 tables of 8 and 8 tables of 12
46 tables of 8 and 6 tables of 12
49 tables of 8 and 4 tables of 12
52 tables of 8 and 2 tables of 12
55 tables of 8 and 0 tables of 12

What's the Problem?

Boxes of 30	Boxes of 40	Boxes of 80
12	0	0
8	1	1
8	3	0
4	0	3
4	2	2
4	4	1
0	9	0
0	7	1
0	5	2
0	3	3
0	1	4

Page 2

The Puzzler
36·8 ÷ 4 = 9·2
39·9 ÷ 7 = 5·7
46·8 ÷ 6 = 7·8

What's the Problem?
5 cars, 1 bicycle
4 cars, 3 bicycles
3 cars, 5 bicycles
2 cars, 7 bicycles
1 car, 9 bicycles
11 bicycles

As before plus:
2 tricycles, 4 cars
2 tricycles, 3 cars, 2 bicycles
2 tricycles, 2 cars, 4 bicycles
2 tricycles, 1 car, 6 bicycles
2 tricycles, 8 bicycles
4 tricycles, 2 cars, 1 bicycle
4 tricycles, 1 car, 3 bicycles
4 tricycles, 5 bicycles
6 tricycles, 1 car
6 tricycles, 2 bicycles

Let's Investigate

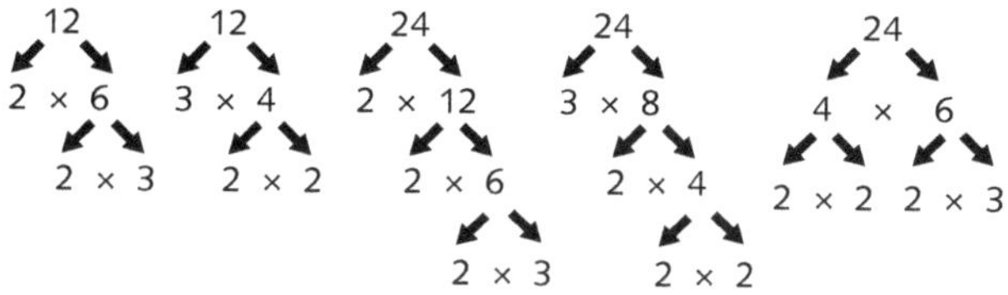

Factor trees for other multiples of 12 from 12 to 72 will vary depending on the multiples chosen by the children.

The Language of Maths

43	÷	6	=	7	r 1
(dividend)		(divisor)		(quotient)	(remainder)

Page 3

Looking for Patterns
The square of any whole number divided by 3 either has no remainder or a remainder of 1.

Let's Investigate
Largest answer is 437·15 (874·3 ÷ 2), the smallest is 29·3375 (234·7 ÷ 8)
Largest answer is 43·715 (87·43 ÷ 2), the smallest is 2·93375 (23·47 ÷ 8)

Let's Investigate

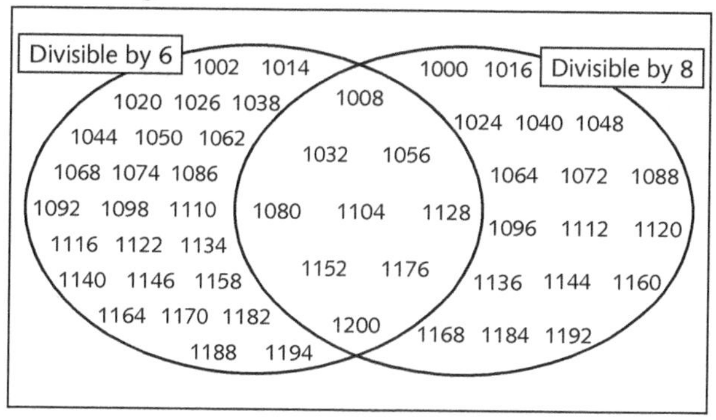

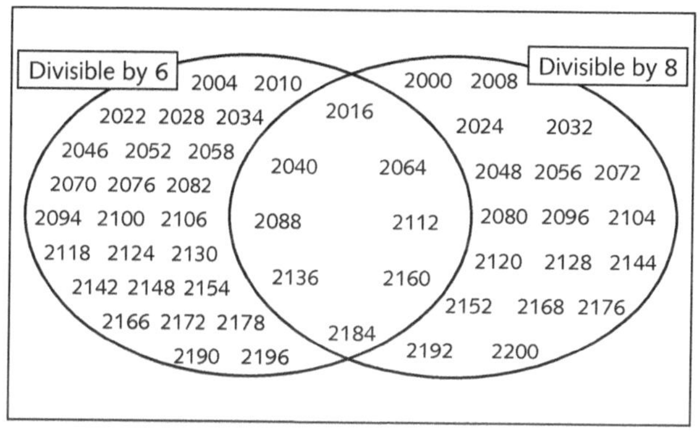

Looking for Patterns

3	3	3	■	■	2	1	4	2
■	5	■	2	1	6	■	9	■
■	9	■	9	■	3	4	8	3
2	7	5	4	■	5	■	2	■
■	7	■	6	■	1	■	9	■
■	5	■	6	2	1	3	4	2

Other answers are possible.

Page 4

Looking for Patterns
6457 61 127 497 002
522 808 357 709 64 283 857
Other answers are possible.

The Puzzler
325 ÷ 25 = 13 (triangular ÷ square = prime)
196 ÷ 7 = 28 (square ÷ prime = triangular)
496 ÷ 31 = 16 (triangular ÷ prime = square)
Other answers are possible.

Inquisitive ant

divisibility
The capacity of one number to be exactly divided by another number.

Issue 9

Mixed operations

Prerequisites for learning

- Understand and use the order of operations
- Calculate mentally with integers
- Use the formal written methods to add, subtract, multiply and divide
- Begin to understand and use simple algebraic conventions
- Use simple expressions and formulae in words and symbols

Resources

pencil and paper
Resource sheet 2: My notes (optional)
Resource sheet 3: Pupil self assessment booklet (optional)
set of dominoes
calculator
computer with Internet access

Teaching support

Page 1

Let's Investigate

- Suggest the children organise their work into a table, e.g.

Size of square	Number of squares
8 × 8 square	1
7 × 7 squares	4
6 × 6 squares	9
5 × 5 squares	16
4 × 4 squares	25
3 × 3 squares	36
2 × 2 squares	49
1 x 1 squares	64

By referring to the table we can see that there are actually 64 + 49 + 36 + 25 + 16 + 9 + 4 + 1 squares on a chessboard: a total of 204.

- Once the children have found the total number of squares, ask them if they can see a pattern in the results (i.e. the square numbers).

Let's Investigate

- Some children will need help in writing an expression or formula to explain what happens. A helpful strategy is to write a formula for each step directly beside it. For example, if n is the first number then,

Step 1: Choose any three-digit number.	n
Step 2: Multiply the number by 10.	$10 \times n$ or $10n$
Step 3: Find the difference between the numbers in Steps 1 and 2.	$10n - n = 9n$
Step 4: Divide the answer from Step 3 by the Step 1 number.	$9n \div n = 9$

- If appropriate, discuss with the children the algebraic convention for omitting the multiplication sign when writing an algebraic expression, i.e. $10 \times n = 10n$

Page 2

Let's Investigate

- Ensure children are familiar with the order of operations.
- As the important aspect of this activity is to think strategically about how to arrange the digits into numbers, and which operations to use in the calculation, allow the children to use a calculator to work out the answers.

The Puzzler

- Given that one possible answer is:
  1000 + 5 + 9 – 2 – 4 – 3 × 6 × 7 × 8 × 1 = 0, you may want to provide some of the children with either the digits or the operations, i.e.

  1000 + 5 □ 9 □ 2 □ 4 □ 3 □ 6 □ 7 □ 8 □ 1 = 0

  or

  1000 + □ + □ – □ – □ – □ × □ × □ × □ × □ = 0

Let's Investigate

- This activity assumes children will use a traditional set of 28 'Double 6 Dominoes'. If appropriate, ask the children to investigate what calculations they could make using a set of 55 'Double 9 Dominoes'.

Page 3

Money Matters

- Costs of keeping a pet include:
  - Equipment
  - Food
  - Vet
  - Insurance
  - Toys

The Puzzler

- Ensure the children understand how the squares work.
- What do the children notice about the diagonals in the squares?

Answers have a difference of 3 (the difference between 5 and 8)

Answers have a difference of 13 (the sum of 5 and 8)

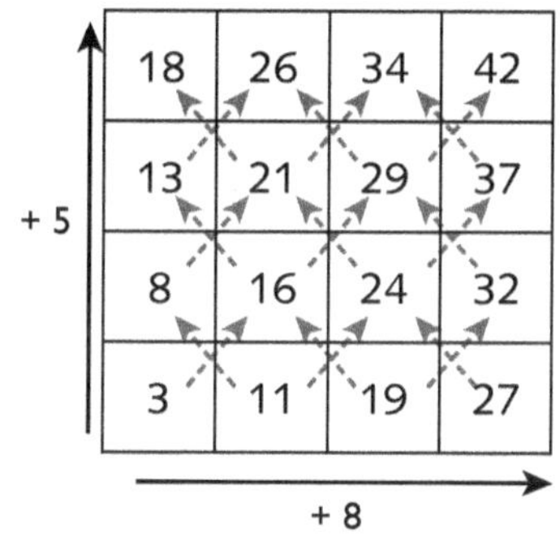

Page 4

Let's Investigate

- Children investigate other patterns and relationships using a calendar.

Let's Investigate

- Children investigate other patterns and relationships using a calendar.

The Puzzler

- 2 large buttons + 4 small buttons = 80p
  2 large buttons + 3 small buttons = 69p (subtract)
  1 small button = 11p

  1 small button costs 11p.
  Therefore two large buttons cost 80p – 44p = 36p.
  Therefore the cost of one large button is 18p.

AfL

- How did you work out the answer to this problem? How did you begin?
- What patterns did you notice? Can you explain this in words / using symbols?
- What was the rule you needed to apply in order to solve this problem?
- What calculations were you able to make? What things did you need to look out for?

Answers

Page 1

Let's Investigate
204 squares

Let's Investigate
The answer is always 9.
Expression may be similar to the following:
If n is the first number then:
Step 1: n
Step 2: $10 \times n$ or $10n$
Step 3: $10n - n = 9n$
Step 4: $9n \div n = 9$

Page 2

Let's Investigate
Results of the investigation will vary.
The smallest number is 1 ÷ 23 456 789 = 0·0000000426
The largest number is 12 345 678 × 9 = 111 111 102

The Puzzler
1000 + 5 + 9 – 2 – 4 – 3 × 6 × 7 × 8 × 1 = 0
Other answers are possible.

Let's Investigate
Results of the investigation will vary.

Page 3

Money Matters
Results of the investigation will vary.

The Puzzler

+ 5 (upwards)

18	26	34	42
13	21	29	37
8	16	24	32
3	11	19	27

+ 8 (to the right)

– 3 (upwards)

5	11	17	23
8	14	20	26
11	17	23	29
14	20	26	32

+ 6 (to the right)

Page 4

Let's Investigate
Both answers are the same. This works for any month.
Explanation:
If x is the smallest date. Then:
- the date immediately to the right is $x + 1$
- the date directly below is $x + 7$
- the date diagonally below is $x + 8$
- $x + (x + 1) + (x + 7) + (x + 8) = 4x + 16$.

Let's Investigate
The sums of diagonally opposite pairs are the same.
The difference in products of the pairs is 7.
The sum of the five numbers is 75, 5 times the middle number.
This works for any month.
Results of the investigation will vary.

The Puzzler
18p

Inquisitive ant

abacus
A mechanical device for making calculations. It consists of a frame mounted with rods along which beads or balls are moved. The first abaci originated in China around 5000 years ago.

Issue 10

Mixed operations

Prerequisites for learning

- Understand square numbers, cube numbers and other powers
- Understand square roots and use the square-root key on a calculator
- Understand tests of divisibility
- Understand and use the order of operations, including brackets
- Calculate mentally with integers
- Use the formal written methods to add, subtract, multiply and divide

Resources

pencil and paper

Resource sheet 2: My notes (optional)

Resource sheet 3: Pupil self assessment booklet (optional)

calculator

Teaching support

Page 1

Let's Investigate

- Most of the activities in this issue focus on the concept of 'casting out nines'.
- The 'casting out nines' method is a way of checking the answers to addition, subtraction, multiplication and division calculations by looking at the 'digital roots' of all the numbers in the calculation.
- The 'digit sum' (also referred to as 'digital sum') of a given integer is the sum of all its digits. For example, the digit sum of 3679 is calculated as $3 + 6 + 7 + 9 = 25$.
- The 'digital root' (also referred to as 'repeated digital sum') of a number is the one-digit value obtained by a reoccurring process of summing digits, on each occurrence using the result from the previous occurrence to calculate a 'digit sum'. The process continues until a one-digit number is reached. For example, the digital root of 3679 is 7 because $3 + 6 + 7 + 9 = 25$ and $2 + 5 = 7$.
- This activity introduces children to the concept of 'digital root' as a test of divisibility for nine.

Page 2

Let's Investigate

- Ensure children can find the digital root of any positive whole number, and that they understand how the 'casting out nines' method can be used to check the answer to an addition calculation.
- Encourage children to use the 'casting out nines' method for addition calculations involving four-digit (and larger) numbers.

Let's Investigate

- Ensure children can find the digital root of any positive whole number, and that they understand how the 'casting out nines' method can be used to check the answer to a subtraction calculation.
- Encourage children to use the 'casting out nines' method for subtraction calculations involving four-digit (and larger) numbers.

Page 3

Let's Investigate

- Ensure children can find the digital root of any positive whole number, and that they understand how the 'casting out nines' method can be used to check the answer to a multiplication calculation.
- Encourage children to use the 'casting out nines' method for multiplication calculations involving combinations of two-, three- and four-digit numbers.

Looking for Patterns

- 142 857 × 1 = 142 857
  142 857 × 2 = 285 714
  142 857 × 3 = 428 571
  142 857 × 4 = 571 428
  142 857 × 5 = 714 285
  142 857 × 6 = 857 142

The pattern can be more easily identified if the six digits are written round a circle or hexagon, i.e.

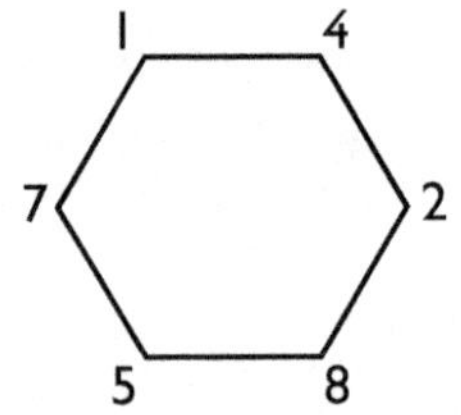

Page 4

Let's Investigate

- Ensure children can find the digital root of any positive whole number, and that they understand how the 'casting out nines' method can be used to check the answer to a division calculation.
- Encourage children to use the 'casting out nines' method for division calculations involving combinations of two-, three- and four-digit numbers.

Let's Investigate

- What if children use three-digit numbers?

AfL

- Explain to me the 'casting out nines' method.
- How do you apply the 'casting out nines' method to answering addition / subtraction / multiplication / division calculations?
- Do you like the 'casting out nines' method? Is this a method you might ever use? Why? / Why not?
- What patterns did you notice?

Answers

Page 1

Let's Investigate
Results of the investigation will vary.

Page 2

Let's Investigate
Results of the investigation will vary.

Let's Investigate
Results of the investigation will vary.

Page 3

Let's Investigate
Results of the investigation will vary.

Looking for Patterns
When 142 857 is multiplied by 1, 2, 3, 4, 5 or 6, the products are the same digits in the same order, but starting in a different place each time.

Page 4

Let's Investigate
Results of the investigation will vary.

Let's Investigate
Numbers less than 40 all make chains with six repeating links, except for 13, 26 and 39 which have a one-link repeating chain. Numbers larger than 40 also form chains. However, the results of what happens will vary depending on the numbers children choose and the rules they apply.

Inquisitive ant

arithmetic
The area of mathematics that deals with calculating numbers, i.e. addition, subtraction, multiplication and division.

Issue 11

Mixed operations

Prerequisites for learning

- Understand square numbers, cube numbers and other powers
- Understand square roots and use the square-root key on a calculator
- Understand tests of divisibility
- Understand and use the order of operations, including brackets
- Calculate mentally with integers
- Use the formal written methods to add, subtract, multiply and divide

Resources

pencil and paper

Resource sheet 2: My notes (optional)

Resource sheet 3: Pupil self assessment booklet (optional)

calculator

set of 0–9 digit cards, addition, subtraction, multiplication, division and equals symbol cards, three decimal point cards (optional)

collection of different breakfast cereals (optional)

computer with Internet access

Teaching support

Page 1

The Puzzler

- Children can create their own mazes on a computer. These can then be displayed on an interactive whiteboard (IWB) for the whole class to solve. When solving the mazes on the IWB, children do not need to copy the maze, they can just write the calculation as a long string of numbers and operations, i.e. $6 \times 3 = 18 - 9 = 9 + 3 = 12 \div 4 = 3 + 5 = 8 + 2 = 10 \div 5 = 2$.

Money Matters

- Children will need to use the Internet to find out the information necessary to complete this activity.
- The important aspect of this activity is for children to be able to give justifiable reasons as to why some things are more expensive now than they were in 1970, and why other things are now cheaper.

Page 2

Let's Investigate

- Ensure children realise that they can only use each digit once in each calculation, i.e.

Allowed | 9 | · | 4 | – | 3 | · | 6 | = | 5 | · | 8 |

Not allowed | 7 | · | 9 | – | 4 | · | 5 | = | 3 | · | 4 |

- Suggest children use a set of 0–9 digit cards, addition, subtraction, multiplication, division and equals symbol cards, and three decimal point cards.

Money Matters

- The average weekly salary in 1970 was £22.
- The average weekly salary in 2010 was £490.

At Home

- You may wish the children to undertake this activity in school rather than at home. If so, provide a collection of different breakfast cereals.
- If the children do undertake this activity at home, once they have completed the task ensure that there is an opportunity in class for pairs or groups of children to discuss their results.

Page 3

The Puzzler

- Ensure children understand and can use squares, cubes and other powers, square roots, and the order of operations, including brackets.

Sports Update

- a. A drop goal plus a try with a conversion is worth ten points.
  b. The team scored two more drop goals than tries with conversions, together these were worth six points.
- If the team scored six lots of a. (drop goal plus a try with a conversion), this is six lots of ten points, which is a total of 60 points.
  In addition to this they scored b. (two more drop goals which is six points).
  Together this makes 66 points.
  The team therefore scored six tries with conversion and 6 + 2 = 8 drop goals.

The Puzzler

- None of the digits can be zero.
- One strategy for finding a solution to the puzzle is to systematically write down all the possible combinations of three digits whose sum is 16, and then work out the products, i.e.

9, 6, 1 (54)	8, 7, 1 (56)	7, 7, 2 (98)	6, 6, 4 (144)
9, 5, 2 (90)	8, 6, 2 (96)	7, 6, 3 (126)	6, 5, 5 (150)
9, 4, 3 (108)	8, 5, 3 (120)	7, 5, 4 (140)	
	8, 4, 4 (128)		

The 8 must be the last digit as the number is a multiple of 2.

Page 4

Sports Update

- There is no one correct order of how difficult the workouts are. The important aspect of this activity is the reasoning children give for their ranking of the workouts.
- Explain that as each workout is 30 mins long, each of the ten columns on each of the daily programmes represents 3 mins.

The Puzzler

- There are 100 numbers, or 50 subtraction calculations each with an answer of 1. Therefore the sum of 50 1s is 50.
- There are 100 numbers, or 50 subtraction calculations each with an answer of 1. Therefore the product of 50 1s is 1.

AfL

- What strategies did you use to answer this problem / puzzle?
- What does the term 'cost of living' mean? What did you find out about it?
- What rules / patterns did you use to help you answer this problem / puzzle?
- How did trial and improvement help with this problem / puzzle?

Answers

Page 1

The Puzzler

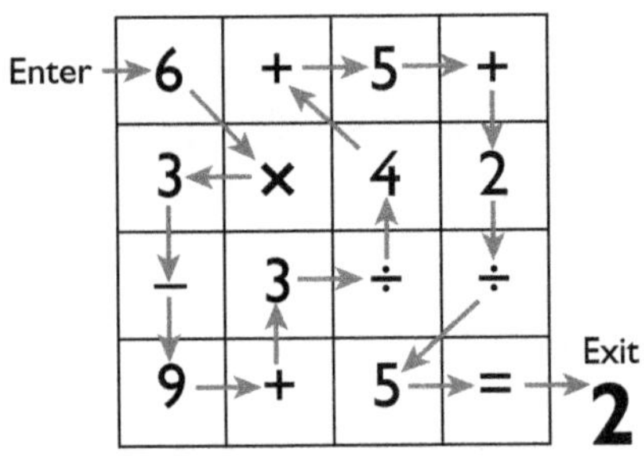

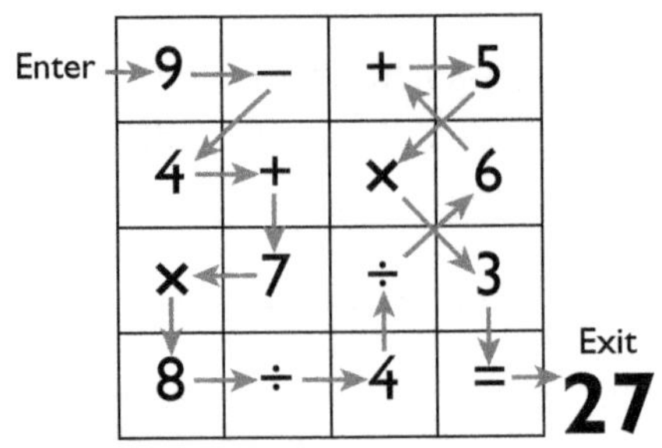

Money Matters
Answers will vary.

Page 2

Let's Investigate
Results of the investigation will vary.

Money Matters
Results of the investigation will vary.

At Home
Results of the investigation will vary.

Page 3

The Puzzler
It is possible to make all the numbers from 0 to 30.
Calculations will vary.

Sports Update
8 drop goals and 6 converted tries.

The Puzzler
358 or 538

Page 4

Sports Update

	Level 1	Level 2	Level 3	Level 4	Level 5
Monday (Programme 3)	6 min	6 min	–	–	18 min
Tuesday (Programme 5)	–	–	15 min	–	15 min
Wednesday (Programme 7)	12 min	–	9 min	–	9 min
Thursday (Programme 1)	–	15 min	–	9 min	6 min
Friday (Programme 4)	–	12 min	–	–	18 min
Saturday (Programme 2)	–	12 min	–	12 min	6 min
Sunday (Programme 6)	6 min	6 min	6 min	6 min	6 min

Ranking and justifications will vary.

The Puzzler
50
1

Inquisitive ant

BODMAS
The order of operations, i.e. B (brackets) O (power of, e.g. 4^2) D (division) M (multiplication) A (addition) S (subtraction).

Issue 12

Mixed operations

Prerequisites for learning

- Identify multiples and factors, including finding all factor pairs of a number
- Calculate mentally with integers
- Use formal written methods to add, subtract, multiply and divide
- Read Roman numerals to at least 1000 (M)
- Solve problems which require answers to be rounded to specified degrees of accuracy
- Estimate the answer to a calculation

Resources

pencil and paper
Resource sheet 2: My notes (optional)
Resource sheet 3: Pupil self assessment booklet (optional)
calculator
computer with Internet access

Teaching support

Page 1

The Puzzler

- Once the children have completed the activity ask them to find out what is the smallest answer that can be made by arranging the numbers cards into two two-digit numbers.
- Ask the children to choose four other digits to arrange into two two-digit numbers. What generalisations can they make?
- What is the largest and smallest number the children can make by arranging the digits 1, 2, 3, 4, 5 and 6 into two three-digit numbers and multiplying them together?

Sports Update

- The lowest common multiple (LCM) of 6 and 8 is 24. So Ethan is first lapped after 24 minutes. In this time he has run three laps of the running track (24 ÷ 8 = 3).
- If necessary, suggest the children draw a diagram to show where Ethan and Joel are on the running track after 6, 12, 18 and 24 minutes.

Money Matters

- Charlie bought three wraps for £3.60. So one wrap costs £1.20. Louise bought one wrap and two cakes for £3, so one cake must cost 90p (£3 – £1.20 = £1.80 / £1.80 ÷ 2 = 90p). Natalie bought a sandwich, a cake and a can of drink for £3.90, so the combined cost of a sandwich and a can of drink must be £3 (£3.90 – 90p = £3). Therefore, the cost of Will buying a sandwich, a wrap, and a can of drink is £4.20 (£3 + £1.20 = £4.20).

Page 2

Technology Today

- Allow the children to use a calculator to help them work out the mistakes that Harry made.
- Ask the children to work out the correct answer to each calculation, using a calculator or not – whichever is appropriate.

Focus on Science

- As this question involves dividing by nine, begin by asking the children to divide $2\frac{1}{2}$ million by ten to get a less accurate approximation. Then encourage them to use the formal written method (as opposed to using a calculator) to work out a more accurate approximation.
- If necessary, remind the children of the need to round to an appropriate degree of accuracy.

Looking for Patterns

- A number is divisible by nine if the sum of its digits is a multiple of nine. The ten known digits add up to 44 (i.e. 0 + 7 + 9 + 3 + 2 + 4 + 6 + 2 + 3 + 8). So the missing digit must be 1 (44 + 1 = 45, and 45 is a multiple of 9).

Page 3

The Puzzler

- Ensure the children are able to read Roman numerals to at least 1000 (M) and use their knowledge of Roman numerals to work out Roman numerals up to 5000.
- If necessary, tell the children some of the calculations using Hindu-Arabic numerals.

Across	Down
1 103 – 56	1 2651 + 1677
2 374 + 465	2 7 × 12
4 3 × 8	3 803 – 432
6 190 ÷ 5	4 168 ÷ 8
7 19 × 25	5 4000 + 85
8 4416 + 3905	9 260 – 4
10 4238 – 2843	11 40 x 8
12 2431 × 2	12 1907 + 2222
13 3000 + 4001	14 3286 – 1793
16 104 ÷ 4	15 51 × 5
18 1000 – 41	17 16 × 4
20 17 × 50	19 710 ÷ 10

Page 4

Famous Mathematicians

- 28 is the only two-digit perfect number. 1, 2, 4, 7, 14 and 28 are the factors of 28 and, 1 + 2 + 4 + 7 + 14 = 28.
- Tell the children that there is only one three-digit perfect number (496). Can they work it out? If necessary give the children some clues, i.e. it is less than 500.

The Puzzler

- The ages of the five people in the Wheeler family total 111 years.
- Trevor and Sally total 55 years, and Helen and Charlie total 45 years. This means that the combined age of Trevor, Sally, Helen and Charlie is 100 years old (55 + 45 = 100). Therefore, Nathan is 11 years old (111 – 100 = 11).
- So:
  - If Nathan is 11 years old, then Sally is 13 years old (24 – 11 = 13)
  - If Nathan is 11 years old, then Helen is 39 years old (50 – 11 = 39).
  - If Helen is 39 years old, then Charlie is 6 years old (45 – 39 = 6).
  - If Sally is 13 years old, then Trevor is 42 years old (55 – 13 = 42).

Focus on Science

- Allow the children to use a calculator for this activity. The most important aspect of this activity is the explanation that children give in working out the maximum distance that each plane can fly on a full tank of fuel.
- Given that a Boeing 777 uses 7855 litres of fuel per hour and that it holds a maximum of 181 280 litres of fuel, a Boeing 777 can fly approximately 23 hours (181 280 ÷ 7855 = 23·078…) So, if a Boeing 777 can travel 590 miles per hour, the approximate maximum distance that a Boeing 777 can fly on a full tank of fuel is 13 617 miles (590 × 23·08 ≈ 13 617).
- Given that a Boeing 747 uses 15 057 litres of fuel per hour and that it holds a maximum of 216 840 litres of fuel, a Boeing 747 can fly approximately 14 hours (216 840 ÷ 15 057 = 14·401…) So, if a Boeing 747 can travel 567 miles per hour, the approximate maximum distance that a Boeing 747 can fly on a full tank of fuel is 8165 miles (567 × 14·4 ≈ 8165).
- Given that a Boeing 737 uses 2965 litres of fuel per hour and that it holds a maximum of 26 020 litres of fuel, a Boeing 737 can fly approximately 23 hours (26 020 ÷ 2965 = 8·775…) So, if a Boeing 737 can travel 511 miles per hour, the approximate maximum distance that a Boeing 737 can fly on a full tank of fuel is 4487 miles (511 × 8·78 ≈ 4487).
- When the children have completed the activity ask them to find out the furthest destination that each plane can fly to, from a UK airport, without refuelling.

AfL

- How did you work out the answer to this problem/puzzle? How did you begin?
- Did you draw a diagram to help you work out the answer? How did it help? Explain your diagram to me.
- What patterns did you notice? Can you explain this to me?
- Did you make any approximations? Explain to me what you did.
- What patterns/relationships did you notice? How did these help you in solving this problem/puzzle?
- What calculations did you need to make? What things did you need to look out for?

Answers

Page 1

The Puzzler
41 × 32 = 1312

Sports Update
3 laps

Money Matters
Will's lunch was £4.20

Page 2

Technology Today
36·5 + 4·78 = 84·3 ✘
Harry inputted 36·5 + 47·8 (decimal point in the wrong position)

234 × 15 = 3645 ✘
Harry inputted 243 × 15 (reversed the tens and ones digits in the number 234)

6214 – 3506 = 5858 ✘
Harry inputted 6214 – 356 (omitted the tens digit, i.e. 0, from the number 3506)

684 ÷ 9 = 85·5 ✘
Harry inputted 684 ÷ 8 (number 8 instead of number 9)

4531 + 1875 = 2656 ✘
Harry inputted 4531 – 1875 (subtraction symbol instead of addition symbol)

Focus on Science
2 500 000 ÷ 9 = $277\ 777{\cdot}\dot{7}$
So approximately 277 778 flushes.

Looking for Patterns
The garage's mobile number is 07932 462 381.

Page 3

The Puzzler

[1] 4	7	■	[2] 8	[3] 3	9	■	[4] 2	[5] 4
[6] 3	8	■	[7] 4	7	5	■	1	0
2	■	■	■	1	■	■	■	8
[8] 8	3	[9] 2	1	■	[10] 1	[11] 3	9	5
■	■	5	■	■	■	2	■	■
[12] 4	8	6	2	■	[13] 7	0	0	[14] 1
1	■	■	■	[15] 2	■	■	■	4
[16] 2	[17] 6	■	[18] 9	5	9	■	[19] 7	9
9	4	■	[20] 8	5	0	■	1	3

Page 4

Famous Mathematicians
28 is the only two-digit perfect number.

The Puzzler
Nathan is 11 years old.
Sally is 13 years old.
Helen is 39 years old.
Charlie is 6 years old.
Trevor is 42 years old.

Focus on Science
Boeing 777: Accept any answer in the range 13 610 miles – 13 620 miles.
Boeing 747: Accept any answer in the range 8160 miles – 8170 miles.
Boeing 737: Accept any answer in the range 4480 miles – 4490 miles.

Inquisitive ant

reasoning
To think about something in a logical and sensible way.

Issue 13

Fractions

Prerequisites for learning

- Identify prime numbers
- Identify, name and write equivalent fractions
- Use common factors to simplify fractions
- Compare and order fractions
- Express a smaller whole number as a fraction of a larger one
- Relate fractions to multiplication and division

Resources

pencil and paper
Resource sheet 2: My notes (optional)
Resource sheet 3: Pupil self assessment booklet (optional)
Resource sheet 19: 2 cm squared paper
A4 paper
ruler
20 different items, each displaying where it was manufactured
computer with Internet access

Teaching support

Page 1

Around the World

- Prior to the activity, gather together 20 different items each displaying where it was manufactured. Alternatively, ask each child to bring in a few items or packaging from home and, from the combined items, each child can then choose 20 items to investigate.

The Arts Roundup

- If a child has no musical knowledge, ask them to work with someone in the class who plays a musical instrument.

Let's Investigate

- Suggest children try other rectangles with multiples of six squares in them.

Page 2

The Puzzler

- Suggest children look at equivalent fractions of $\frac{1}{4}$, $\frac{2}{5}$ and $\frac{1}{2}$.

Let's Investigate

- One game could be that children take turns to roll two 1–9 dice and use the numbers to make a proper fraction. They then write this on the number line. The first child to get three of their fractions together, without an opponent's fraction in between, wins.

The Puzzler

- Ask the children to investigate the patterns and relationships that occur.

Page 3

Money Matters

- Go through an example with the children:

  £14 and £112
  $= \frac{14}{112}$
  $= \frac{1}{8}$
  £14 is $\frac{1}{8}$ of £112.

Let's Investigate

- Provide the children with a list of all the prime numbers from 3 to 19, i.e. 3, 5, 7, 11, 13, 17 and 19.

What's the Problem?

- Discuss with the children the need to find a common denominator, e.g. $\frac{1}{2} = \frac{3}{6}$ and $\frac{1}{3} = \frac{2}{6}$.
  However, this leaves no fraction in between.
  Suggest the children multiply each fraction by two.
  So, $\frac{2}{6} = \frac{4}{12}$ and $\frac{3}{6} = \frac{6}{12}$. Leaving $\frac{5}{12}$ to fall between $\frac{1}{2}$ and $\frac{1}{3}$.
- Ask the children to find more than one answer.

Let's Investigate

- Encourage children to start with the simplest unit fractions and then make as many equivalents as they can i.e. $\frac{1}{2} = \frac{2}{4} = \frac{3}{6} = \frac{4}{8}$ etc. They can then pick pairs that do not use the same numbers.

Page 4

What's the Problem?

- Tell the children you are thinking of a number; half this number is ten. Ask the children what your number is (20). Ask them how they knew this (multiplied 10 by 2). Write this down.

$\square \times \frac{1}{2} = 10$

$\square = 10 \times \frac{2}{1} = 20$

If you want to work back from the answer, you multiply it by each fraction inverted.

The Puzzler

- Mr and Mrs Lane = $\frac{2}{3}$ × 120 years = 80 years
  Combined age of twins = $\frac{1}{5}$ × 120 years = 24 years
  ∴ Age of twins = 12 years
  ∴ Third child = 120 years – 80 years – 24 years = 16 years.
  Sum of children's ages = $\frac{8}{9}$ × Mr Lane's age
  ∴ Age of Mr Lane = $\frac{9}{8}$ × 40 years = 45 years
  ∴ Age of Mrs Lane = 80 years – 45 years = 35 years.

Looking for Patterns

- The front page is $\frac{1}{36}$ of the total pages and is $\frac{9}{10}$ news therefore $\frac{1}{36} \times \frac{9}{10}$ of the paper is news
  Pages 2–5 are $\frac{4}{36}$ of the total pages and are $\frac{8}{10}$ news therefore $\frac{4}{36} \times \frac{8}{10}$ of the paper is news
  Pages 6–9 are $\frac{4}{36}$ of the total pages and are $\frac{7}{10}$ news therefore $\frac{4}{36} \times \frac{7}{10}$ of the paper is news
  Pages 10–13 are $\frac{4}{36}$ of the total pages and are $\frac{6}{10}$ news therefore $\frac{4}{36} \times \frac{6}{10}$ of the paper is news
  Pages 14–35 are $\frac{22}{36}$ of the total pages and are $\frac{5}{10}$ news therefore $\frac{22}{36} \times \frac{5}{10}$ of the paper is news
  The back page is $\frac{1}{36}$ of the total pages and is $\frac{9}{10}$ news therefore $\frac{1}{36} \times \frac{9}{10}$ of the paper is news.
  Therefore the total fraction of the paper that is news is $(\frac{1}{36} \times \frac{9}{10}) + (\frac{4}{36} \times \frac{8}{10}) + (\frac{4}{36} \times \frac{7}{10}) + \ldots$

AfL

- What do these numbers mean?
- What patterns did you recognise?
- How did you work out that fraction? What clues did you use? How did they help?
- How did you work out the solution to this puzzle?
- Explain to me how you worked out the answer.

Answers

Page 1

Around the World
Answers will vary.

The Arts Roundup
They are music time signatures. The bottom number represents the note length as a fraction of a whole note, and the top number represents how many of these notes there are in one bar of music.

$\frac{2}{4}$ equals two $\frac{1}{4}$ notes (crotchets) in a bar
$\frac{3}{4}$ equals three $\frac{1}{4}$ notes in a bar
$\frac{6}{8}$ equals six $\frac{1}{8}$ notes (quavers) in a bar.

Let's Investigate
Shadings will vary. However, the whole rectangle is shaded:

$\frac{1}{2}+\frac{1}{4}+\frac{1}{6}+\frac{1}{12}=1.$

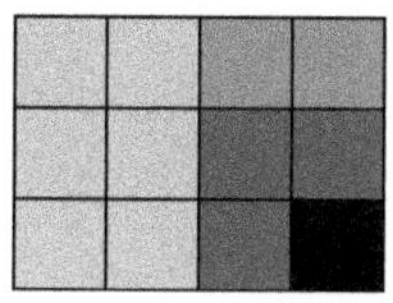

Results of the investigation will vary.

Some of the simpler rectangles are:

Odd multiples of 6 squares: $\frac{1}{2}+\frac{1}{3}+\frac{1}{6}$

Even multiples of 6 squares: $\frac{1}{2}+\frac{1}{4}+\frac{1}{6}+\frac{1}{12}$

Multiples of 20 squares: $\frac{1}{2}+\frac{1}{4}+\frac{1}{5}+\frac{1}{20}$

Page 2

The Puzzler

$\frac{1\cancel{6}}{\cancel{6}4}=\frac{1}{4}, \frac{2\cancel{6}}{\cancel{6}5}=\frac{2}{5}, \frac{4\cancel{9}}{\cancel{9}8}=\frac{4}{8}=\frac{1}{2}.$

Let's Investigate
Games will vary.

The Puzzler
The A4 sheet of paper has been divided into 16 pieces.
Each small folded piece represents $\frac{1}{16}$ of the A4 sheet.

If the sheet of paper were folded twice more it would be divided into 64 pieces.
Each small folded piece would represent $\frac{1}{64}$ of the A4 sheet.

Page 3

Money Matters

$\frac{1}{9}$	$\frac{1}{14}$	$\frac{1}{15}$
$\frac{1}{39}$	$\frac{1}{14}$	$\frac{1}{32}$
$\frac{1}{12}$	$\frac{1}{47}$	$\frac{1}{87}$
$\frac{1}{16}$	$\frac{1}{27}$	$\frac{1}{23}$

Let's Investigate

$\frac{3+11}{7}=\frac{14}{7}=2$

$\frac{3+17}{5}=\frac{20}{5}=4$

$\frac{5+7}{3}=\frac{12}{3}=4$

$\frac{5+13}{3}=\frac{18}{3}=6$

$\frac{5+19}{3}=\frac{24}{3}=8$

$\frac{7+11}{3}=\frac{18}{3}=6$

$\frac{7+13}{5}=\frac{20}{5}=4$

$\frac{7+17}{3}=\frac{24}{3}=8$

$\frac{7+19}{13}=\frac{26}{13}=2$

$\frac{11+13}{3}=\frac{24}{3}=8$

$\frac{11+17}{7}=\frac{28}{7}=4$

$\frac{11+19}{3}=\frac{30}{3}=10$

$\frac{11+19}{5}=\frac{30}{5}=6$

$\frac{13+17}{5}=\frac{30}{5}=6$

$\frac{17+19}{3}=\frac{36}{3}=12$

The sum of two primes divided by a different prime always gives an even answer.

What's the Problem?

Many answers are possible, e.g. $\frac{5}{12}, \frac{9}{24}, \frac{11}{24}, \frac{7}{18}, \frac{4}{9}$.

Let's Investigate
Using four numbers from 1 to 10: variations of

$\frac{1}{2}=\frac{3}{6}=\frac{48}{96}=\frac{5}{10}$
$\frac{1}{3}=\frac{2}{6}=\frac{3}{9}$
$\frac{2}{3}=\frac{4}{6}=\frac{6}{9}$
$\frac{1}{4}=\frac{2}{8}$
$\frac{3}{4}=\frac{6}{8}$
$\frac{1}{5}=\frac{2}{10}$

Using six numbers from 1 to 20: variations of

$\frac{1}{2}=\frac{3}{6}=\frac{4}{8}=\frac{5}{10}=\frac{6}{12}=\frac{7}{14}=\frac{8}{16}=\frac{9}{18}=\frac{10}{20}$
$\frac{1}{3}=\frac{2}{6}=\frac{3}{9}=\frac{4}{12}=\frac{5}{15}=\frac{6}{18}$
$\frac{2}{3}=\frac{4}{6}=\frac{6}{9}=\frac{8}{12}=\frac{10}{15}=\frac{12}{18}$
$\frac{1}{4}=\frac{2}{8}=\frac{3}{12}=\frac{4}{16}=\frac{5}{20}, \frac{3}{4}=\frac{6}{8}=\frac{9}{12}=\frac{12}{16}=\frac{15}{20}$
$\frac{1}{5}=\frac{2}{10}=\frac{3}{15}=\frac{4}{20}, \frac{2}{5}=\frac{4}{10}=\frac{6}{15}=\frac{8}{20}$
$\frac{3}{5}=\frac{6}{10}=\frac{9}{15}=\frac{12}{20}, \frac{4}{5}=\frac{8}{10}=\frac{12}{15}=\frac{16}{20}$
$\frac{1}{6}=\frac{2}{12}=\frac{3}{18}, \frac{5}{6}=\frac{10}{12}=\frac{15}{18}$

Page 4

What's the Problem?
1200

The Puzzler
Mr Lane is 45 years old.
Mrs Lane is 35 years old.
The twins are 12 years old.
The other child is 16 years old.

Looking for Patterns

$\frac{53}{90}$ of the paper is news.

Inquisitive ant

common fraction (also referred to as a *simple fraction* or a *vulgar fraction*)
A fraction that consists of two whole numbers separated by a horizontal (or slanting) line, e.g. $\frac{1}{2}$ or $\frac{3}{4}$, as opposed to a decimal fraction, e.g. 0·5 or 0·75.

Issue 14

Fractions

Prerequisites for learning

- Identify, name and write equivalent fractions
- Recognise mixed numbers
- Use common factors to simplify fractions
- Add and subtract fractions
- Connect multiplication by a fraction to using fractions as operators (fractions of), and to division.
- Find fractions of numbers and quantities
- Multiply simple pairs of proper fractions
- Use mental methods of calculation with fractions

Resources

pencil and paper
Resource sheet 2: My notes (optional)
Resource sheet 3: Pupil self assessment booklet (optional)
Resource sheet 5: Music paper
dozen sized egg carton and 12 counters, or similar (optional)
set of 0–9 digit cards, a multiplication symbol card and an equals symbol card (optional)
calculator
1–6 dice

Teaching support

Page 1

Let's Investigate

- This activity provides a practical approach to the addition of fractions.
- Encourage children to reduce the fractions to their simplest form.
- Provide the children with a dozen sized egg carton and 12 counters, or similar.

Let's Investigate

- This activity provides a practical approach to multiplying pairs of fractions and seeing the relationship between fractions, multiplication and division.
- Provide the children with a dozen sized egg carton and 12 counters, or similar.

Page 2

The Puzzler

- Ensure children understand how an addition fraction wall works.
- Children make subtraction fraction walls.

Let's Investigate

- Ensure children realise that they can only use each digit once in each calculation, i.e.

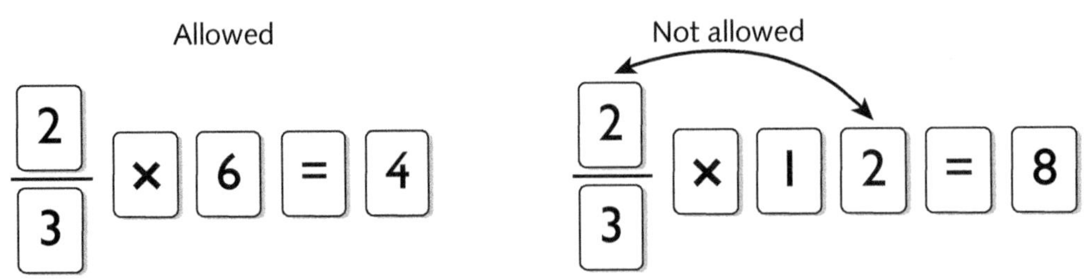

- Suggest children use a set of 0–9 digit cards, a multiplication symbol card and an equals symbol card.

Page 3

The Arts Roundup

- If a child has no musical knowledge, ask them to work with someone in the class who plays a musical instrument.
- What about two whole notes in a bar? What about three whole notes in a bar?

Looking for Patterns

- Ask the children to make up their own addition fraction calculations. Ensure that the children only add pairs of unitary fractions (i.e. with a numerator of 1).

Page 4

Let's Investigate

- Children investigate making answers other than 1, e.g. 0, 2 or 10, by adding or subtracting four different fractions all with the same denominator.

Let's Investigate

- For the first investigation, once children identify the first three-digit number that when divided by eight has a remainder of five, i.e. 101, they should be able to quickly see the other numbers to 150.
- Similarly for the second investigation, once children identify the first three-digit number that when divided by eight has a remainder of seven, i.e. 103, they should be able to quickly see the other numbers to 150.

Let's Investigate

- Reciprocal fractions when multiplied together give an answer of 1, e.g. $\frac{3}{5} \times \frac{5}{3} = 1$.

AfL

- When you were working out the answer to this problem / puzzle did you always think of these numbers as common fractions or did you convert them in your mind to decimal fractions? Did this help? How?
- Explain to me how you add two fractions together.
- What about multiplying two fractions?
- What rules can you tell me for adding / subtracting / multiplying pairs of fractions?
- Did you use trial and improvement to work out the answer? How did this help?

Answers

Page 1

Let's Investigate

1 egg: $\frac{1}{12}$, 2 eggs: $\frac{1}{6}$, 3 eggs: $\frac{1}{4}$, 4 eggs: $\frac{1}{3}$,
5 eggs: $\frac{5}{12}$, 6 eggs: $\frac{1}{2}$, 7 eggs: $\frac{7}{12}$, 8 eggs: $\frac{2}{3}$,
9 eggs: $\frac{3}{4}$, 10 eggs: $\frac{5}{6}$, 11 eggs: $\frac{11}{12}$, 12 eggs: 1

Let's Investigate

A half of $\frac{1}{3}$ is also a $\frac{1}{6}$.
This always happens.

Page 2

The Puzzler

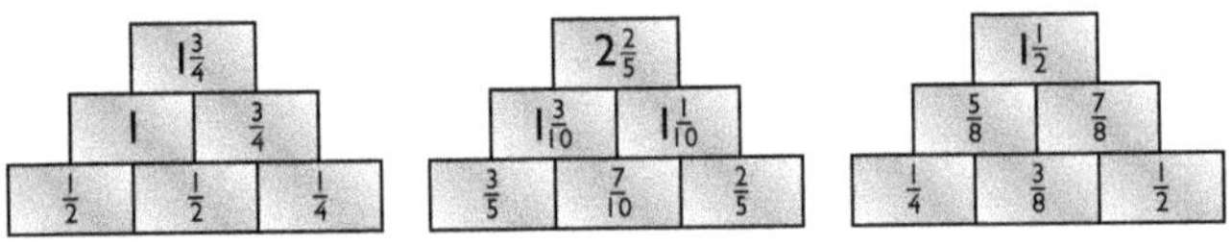

Let's Investigate
Results of the investigation will vary.

Page 3

The Arts Roundup
Results of the investigation will vary.

Looking for Patterns
The pattern for adding unitary fractions is to add the denominators of the two fractions to find the numerator of the answer, and multiply the denominators of the two fractions to find the denominator of the answer.

$\frac{1}{6} + \frac{1}{7} = \frac{13}{42}$ $\frac{1}{7} + \frac{1}{8} = \frac{15}{56}$ $\frac{1}{9} + \frac{1}{10} = \frac{19}{90}$ $\frac{1}{10} + \frac{1}{11} = \frac{21}{110}$

Page 4

Let's Investigate
Results of the investigation will vary.

Let's Investigate

$101 \div 8 = 12\frac{5}{8}$
$109 \div 8 = 13\frac{5}{8}$
$117 \div 8 = 14\frac{5}{8}$
$125 \div 8 = 15\frac{5}{8}$
$133 \div 8 = 16\frac{5}{8}$
$141 \div 8 = 17\frac{5}{8}$
$149 \div 8 = 18\frac{5}{8}$

$103 \div 8 = 12\frac{7}{8}$
$111 \div 8 = 13\frac{7}{8}$
$119 \div 8 = 14\frac{7}{8}$
$127 \div 8 = 15\frac{7}{8}$
$135 \div 8 = 16\frac{7}{8}$
$143 \div 8 = 17\frac{7}{8}$

Let's Investigate
Results of the investigation will vary.

Inquisitive ant

reciprocal fractions
Reciprocal is the word used to describe a number or quality that is related to another by the fact that when multiplied together the product is one. So, *reciprocal fractions* are multiplicative inverses, i.e. $\frac{1}{2} \times \frac{2}{1} = \frac{2}{2} = 1$.

Issue 15

Decimals

Prerequisites for learning

- Understand and use decimal notation for tenths, hundredths and thousandths, and round these numbers
- Relate fractions to their decimal representations and vice versa
- Express a smaller whole number as a fraction of a larger one
- Calculate mentally with integers and decimals
- Use formal written methods to add, subtract and multiply decimals with up to three places
- Use a calculator to solve problems, including those involving decimals

Resources

pencil and paper
Resource sheet 2: My notes (optional)
Resource sheet 3: Pupil self assessment booklet (optional)
0–9 dice
calculator

Teaching support

Page 1

Looking for Patterns

- Encourage children to look for the relationships between the pairs of numbers.

The Puzzler

- Ensure children are familiar with magic squares.
- For the first magic square give children the top left-hand number of 0·6.
- For the second magic square give children the bottom right-hand decimal of 2·2.

Page 2

Looking for Patterns

- Ensure children understand how the table works: that the numbers in the top row refer to the numerator in a fraction, and the numbers in the first column refer to the denominator, i.e. $\frac{?}{2}$ = 0·5.
- It is easier to work out all the numerators and denominators first, before completing the rest of the table.
- Ask the children to create their own table for a friend to solve. However rather than the table showing equivalent fractions and decimals, the table should show equivalent fractions and percentages.

The Puzzler

- Encourage the children to use approximations to estimate the results. For example, in the calculation 34 + 78 = 11·2, given that the whole number in the answer is 11, then the two addends will be 3·4 and 7·8 (3 + 7 = 10).
- Encourage the children to look at the least significant digit in the answer (i.e. the final digit of the answer). Which of the least significant digits of the numbers being added, subtracted or multiplied will result in the least significant digit in the answer?

Page 3

Around the World

- This activity involves children rounding decimals with three decimal places to the nearest whole number, before making comparative statements.

The Puzzler

- Encourage the children to concentrate on getting the tens digit as close as possible first, then the ones digit and then finally the decimals.

Page 4

Money Matters

- Go through the equivalent fractions in the example, i.e.
  $0{\cdot}05 = \frac{5}{100} = \frac{1}{20}$
  $\frac{1}{20}$ of 20p is 1p
  $\frac{5}{100}$ of £10 is 50p.
  Also, if necessary, $0{\cdot}2 = \frac{2}{10} = \frac{1}{5}$ and $0{\cdot}4 = \frac{4}{10} = \frac{2}{5}$

Looking for Patterns

- Ask the children to approximate the answers to $52{\cdot}9 \div 3$, and $207 \div 4{\cdot}1$ and state whether the approximate is more than or less than the actual answer and why.

AfL

- What is the solution to this puzzle? How did you go about finding it out?
- What did you find out? How did you find this out? What rules can you give?
- What patterns did you notice? How could this sequence continue?
- How do you know that you have found all the different possibilities?
- Explain to me how you worked that out. What strategies did you use? How did they help you?
- Tell me one of your statements.

Answers

Page 1

Looking for Patterns
0·2727, 3·6666, 0·2727, 3·6666, 0·2727, 3·6666
0·40909, 2·4444, 0·40909, 2·4444, 0·40909, 2·4444
The answers are the same when the numbers are multiples of each other.

The Puzzler

0·6	0·1	0·8
0·7	0·5	0·3
0·2	0·9	0·4

Magic number is 1·5.

3·4	1·8	3·2
2·6	2·8	3
2·4	3·8	2·2

Magic number is 8·4

Page 2

Looking for Patterns

		numerator				
		1	6	3	2	8
denominator	2	0·5	3	1·5	1	4
	10	0·1	0·6	0·3	0·2	0·8
	5	0·2	1·2	0·6	0·4	1·6
	4	0·25	1·5	0·75	0·5	2
	8	0·125	0·75	0·375	0·25	1

The Puzzler
3·4 + 7·8 = 11·2
7·8 + 9·6 + 14 = 31·4
8·4 – 5·6 = 2·8
43·7 + 8·9 – 26·4 = 26·2
4·6 × 12·8
or 46 × 1·28
or 0·46 × 128 = 58·88
23·5 + 6·12 + 5·15 = 34·77
5·5 + 6·3 + 1·24 – 3·07 = 9·97
84·56 + 3·76 + 1·482 – 35·01 = 54·792

Page 3

Around the World
Statements will vary.

The Puzzler
Answers will vary.

Page 4

Money Matters
5p is worth 0·05 as much as £1.
10p is worth 0·05 as much as £2.
£1 is worth 0·05 as much as £20.

1p is worth 0·2 as much as 5p.
2p is worth 0·2 as much as 10p.
10p is worth 0·2 as much as 50p.
20p is worth 0·2 as much as £1.
£1 is worth 0·2 as much as £5.
£2 is worth 0·2 as much as £10.
£10 is worth 0·2 as much as £50.

2p is worth 0·4 as much as 5p.
20p is worth 0·4 as much as 50p.
£2 is worth 0·4 as much as £5.
£20 is worth 0·4 as much as £50.

2p is worth 0·0004 as much as £50.

Looking for Patterns
$80 \div 1{\cdot}8 = 44{\cdot}\dot{4}$. By rounding up 1·8 to 2, you are dividing 80 by more than if you were dividing 80 by 1·8. Therefore the approximate answer will be less than the actual answer.

Inquisitive ant

≈
Means approximately equal to.

Issue 16

Percentages

Prerequisites for learning

- Recognise the per cent symbol (%) and understand that per cent relates to 'number of parts per hundred'
- Express tenths and hundredths as percentages
- Express one quantity as a percentage of another
- Recognise approximate proportions of a whole and use percentages to describe and compare them
- Understand and interpret pie charts

Resources

pencil and paper
Resource sheet 2: My notes (optional)
Resource sheet 3: Pupil self assessment booklet (optional)
Resource sheet 6: Pie charts
Resource sheet 19: 2 cm squared paper
ruler
concrete apparatus, e.g. interlocking cubes (optional)
computer with Internet access
data handling software

Teaching support

Page 1

Focus on Science

- Children will find this information on the Internet. Be aware that there are various sources and that the percentages may vary slightly.
- Ask the children to create a bar chart of their results using ICT if appropriate.

Let's Investigate

- If necessary, suggest to the children types of numbers for them to investigate, e.g. square numbers, triangular numbers, prime numbers, …

The Language of Maths

- The percentage incline or decline is found by using the following formula $100 \times \frac{\text{rise}}{\text{run}}$, where rise is the height in metres of the top of the slope above the sign and run is the horizontal distance in metres of the top of the slope from the sign.

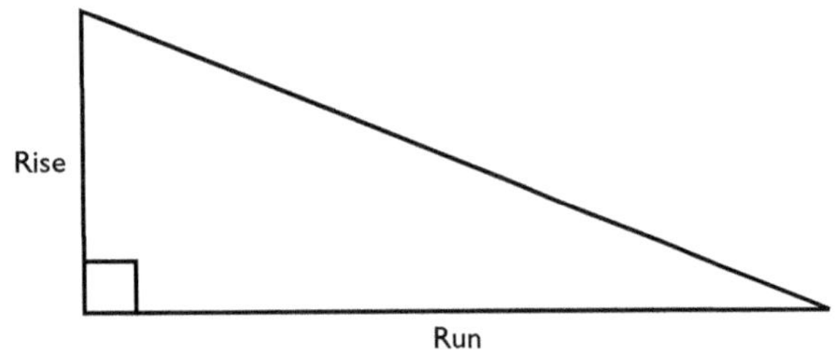

Page 2

The Language of Maths

- Some children may find it difficult to grasp the concept of more than 100%. If this is the case, use concrete apparatus, such as interlocking cubes, to explain it.

What's the Problem?

- The amount of time that either one or the other is in the kitchen alone is:

  the total amount of time that Margot is in the kitchen, minus the amount of time that both Margot and Jerry are in the kitchen together, plus the amount of time Jerry is in the kitchen alone, i.e. 40% – 20% + 10% = 30%.

The Language of Maths

- Some children may find it difficult to grasp the concept of more than 100%. If this is the case, use concrete apparatus, such as interlocking cubes, to explain it.

Page 3

The Arts Roundup

- The purpose of this activity is to consolidate the relationship between fractions and percentages, and to see how the visual representations can be used to make a comparison.

Let's Investigate

- Discuss the second example with the children, ensuring that they understand that 50% of the rectangle is shaded. Encourage the children to give further examples in this way.
- Children could investigate shading 50% of rectangles of different sizes.

Page 4

The Language of Maths

- When constructing pie charts the largest sector should be drawn first, immediately to the right of the 0% division, then the next largest sector and so on around the circle.
- Some children may need help in calculating the proportions of a whole using percentages.

The Language of Maths

- This activity is an extension of the previous activity. You may decide to ask some children to only estimate the percentages of each of the sectors of the pie chart given and not produce their own pie charts.

AfL

- What did you find out? What does it mean? How did you go about this?
- What statements can you tell me?
- What do you notice about your answers?
- What does this mean? Can you give me another example?
- What is a pie chart? What does it show? What might this pie chart be about?
- How did you go about working out the percentage? How would you do this on a calculator?
- What calculation did you do? What does the answer tell you?
- Approximately what percentage of the pie chart is this sector? Why do you think it is that percent?

Answers

Page 1

Focus on Science
Results of the investigation will vary.

Let's Investigate
Results of the investigation will vary.

The Language of Maths
16% decline. 13% incline. 20% incline.

Page 2

The Language of Maths
150% means one and a half times something.
As a decimal it can be expressed as 1·5.
As a fraction it can be expressed in various ways such as $\frac{150}{100}$ or $\frac{3}{2}$.
As a mixed number it can be expressed in various ways such as $1\frac{1}{2}$ or $1\frac{50}{100}$.

What's the Problem?
30%

The Language of Maths
More than 100% is possible. It means that the children are giving all of their normal effort (100%) plus some extra effort.

Page 3

The Arts Roundup

Smith Studios *The Pine Tree*
$\frac{7}{8}$ = 87·5%

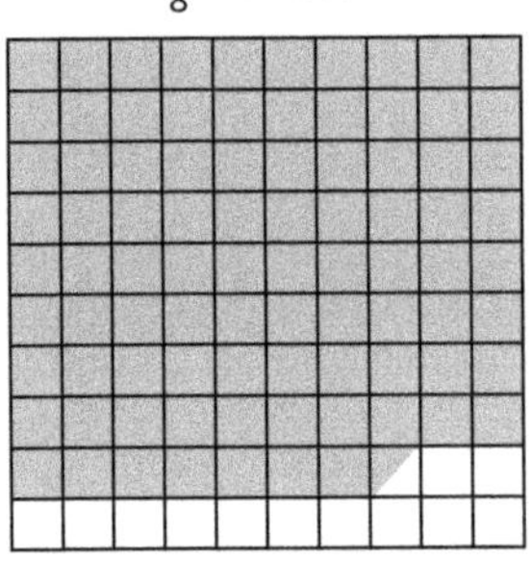

Dench Theatre *Whodunnit*
$\frac{86}{100}$ = 86%

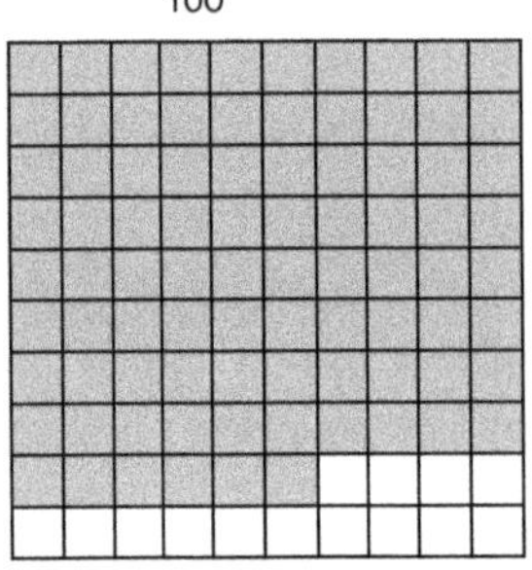

Let's Investigate
Results of the investigation will vary.

Page 4

The Language of Maths

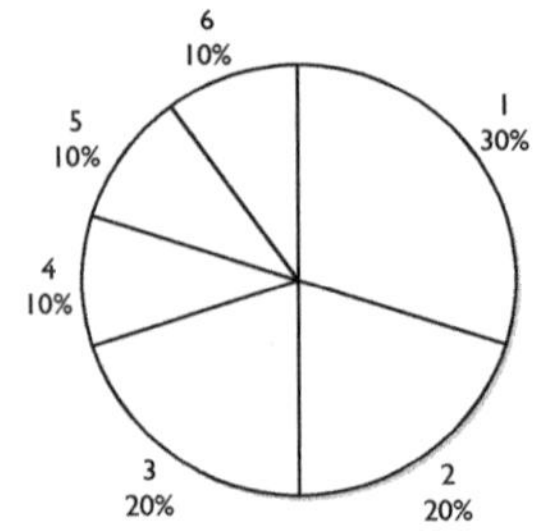

Labels and title will vary.
Pie charts will vary.

The Language of Maths
Estimations will vary slightly from those given below.

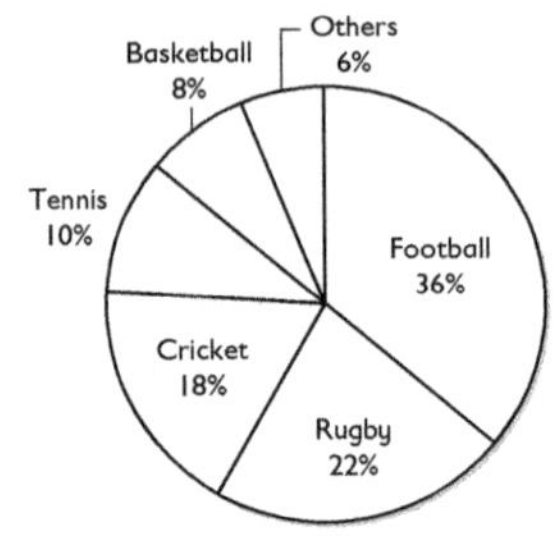

Pie charts will vary.

Inquisitive ant

sector
Part of a circle bounded by two radii and the part of the circumference that lies between them.

Issue 17

Percentages

Prerequisites for learning

- Recognise the per cent symbol (%) and understand that per cent relates to 'number of parts per hundred'
- Recall and use equivalences between fractions, decimals and percentages
- Find percentages of numbers and quantities
- Express one quantity as a percentage of another
- Calculate percentage increases or decreases
- Make and justify estimates and approximations to calculations
- Read and interpret pie charts

Resources

pencil and paper
Resource sheet 2: My notes (optional)
Resource sheet 3: Pupil self assessment booklet (optional)
Resource sheet 6: Pie charts (optional)
calculator
set of 0–9 digit cards, a percentage symbol card, a multiplication symbol card and an equals symbol card (optional)
collection of different breakfast cereals (optional)
computer with Internet access

Teaching support

Page 1

Let's Investigate

- Ensure children realise that they can only use each digit once in each calculation, i.e.

Allowed: 4 0 % × 9 5 = 3 8

Not allowed: 4 0 % × 8 0 = 3 2

- Suggest children use a set of 0–9 digit cards, a percentage symbol card, a multiplication symbol card and an equals symbol card.

Focus on Science

- Ensure children are familiar with pie charts and are able to read and interpret them. Children also need to be able to find percentages of whole number and decimal quantities.

At Home

- You may wish the children to undertake this activity in school rather than at home. If so, provide a collection of different breakfast cereals.
- If the children do undertake this activity at home, once they have completed the task ensure that there is an opportunity in class for pairs or groups of children to discuss their results.

Page 2

Around the World

- Suggest children express all the values in like terms, e.g. percentages, and calculate the total, i.e.

	Autumn Term		Spring	
	1st half	2nd half	1st half	Total
Year 3	20%	30%	10%	60%
Year 4	20%	15%	5%	40%
Year 5	25%	25%	20%	70%
Year 6	10%	10%	10%	30%
				200%

- They then find the average by dividing the overall percentage total by the number of year groups, i.e. four
  200 ÷ 4 = 50%.

What's the Problem?

- Provide the children with a copy of Resource sheet 6: Pie charts, and ask them to draw a pie chart showing what percentage of the different ingredients is in a bag of Natural World's 100% 'All Natural Fruit and Nut Muesli'. You may suggest that they group like ingredients, i.e. all the fruits, together, then all the seeds together and all the nuts together.

Page 3

Sports Update

- Ensure children are familiar with pie charts and are able to read and interpret them. Children also need to be able to find percentages of whole number quantities, and in the case of this activity, quantities involving hour and half hour amounts of time.

Construct

- Children will first need to find the total area of the walls in the classroom and then the area of the walls that is covered by windows.

Page 4

The Arts Roundup

- Remind the children that the percentages given are approximate.

The Arts Roundup

- Remind the children:
  – that they are only expected to find approximate percentages
  – to express all proportions as percentages.

AfL

- Explain this pie chart to me. How does a pie chart represent / compare percentages in a way that is quick and easy to understand?
- How did you work out that percentage?
- What is this expressed as a percentage?
- What did you use to help you work out that answer? Was there a percentage that you already knew that you used to help you? How did this help? Why did you do that?
- How is knowing about the relationship between fractions, decimals and percentages important / helpful?

Answers

Page 1

Let's Investigate
Results of the investigation will vary.

Focus on Science
Heart – 220 ml
Brain – 440 ml
Muscles – 1·375 *l*
Intestines – 825 ml
Kidneys – 1·375 ml
Liver – 550 ml
Other – 715 ml

At Home
Results of the investigation will vary.

Page 2

Around the World
After three half terms, which is half the year, William's Junior School have achieved 50% of their target. At this rate they should raise the £1000 by the end of the school year.

What's the Problem?

	Small 500 g	Medium 1 kg	Large 2 kg
Oat	175 g	350 g	700 g
Wheat	100 g	200 g	400 g
Corn	75 g	150 g	300 g
Dried dates etc	50 g	100 g	200 g
Dried figs etc	25 g	50 g	100 g
Sultanas	25 g	50 g	100 g
Linseed	10 g	20 g	40 g
Sunflower seed	10 g	20 g	40 g
Pecans	10 g	20 g	40 g
Hazelnuts	10 g	20 g	40 g
Almonds	10 g	20 g	40 g

Page 3

Sports Update

My Daily Exercise Routine

	Warm-up	Jogging	Running	Stretching
Monday	4·5 min	7·5 min	15 min	3 min
Tuesday	9 min	15 min	30 min	6 min
Wednesday	4·5 min	7·5 min	15 min	3 min
Thursday	9 min	15 min	30 min	6 min
Friday	4·5 min	7·5 min	15 min	3 min
Saturday	18 min	30 min	60 min	12 min
Sunday	13·5 min	22·5 min	45 min	9 min

Construct
Results of the investigation will vary.
Smaller windows: less heat loss in cold weather and heat gain in hot weather.
Bigger windows: more light.

Page 4

The Arts Roundup
70 String players
18 Brass players
1 Keyboard player
15 Woodwind players
2 Percussion players

The Arts Roundup
Results of the investigation will vary.
An average chamber orchestra has about 50 players.

Inquisitive ant

per
This word means for each or for every thing mentioned. For example, £1.72 per litre or 70 miles per hour.

Issue 18

Fractions, decimals and percentages

Prerequisites for learning

- Convert fractions to their decimal representations
- Recall and use equivalences between fractions, decimals and percentages
- Calculate mentally with integers, fractions, decimals and percentages
- Use formal written methods to add and subtract decimals with up to two places
- Use a calculator to solve problems, including those involving fractions, decimals and percentages

Resources

pencil and paper
Resource sheet 2: My notes (optional)
Resource sheet 3: Pupil self assessment booklet (optional)
calculator
computer

Teaching support

Page 1

The Puzzler

- There are four different paths that give a total of one.

The Language of Maths

- Ensure that the children understand that digits can recur in groups and not just individually.

Page 2

Looking for Patterns

- Children investigate patterns for other numbers and / or calculations.
- If appropriate, the children can create their diagrams using ICT.

Page 3

The Puzzler

- Ensure children can identify equivalent fractions, decimals and percentages.
- Ask the children to create their own fractions, decimals and percentages equivalent dominoes puzzle using 12 or 16 dominoes for a friend to solve.

Looking for Patterns

- This activity involves children using the relationship between decimals and percentages.
- Children write what proportion of each 100 square is not shaded using common fractions reduced to their simplest term.

Page 4

The Puzzler

- Ensure children can identify equivalent fractions, decimals and percentages.
- Once children have completed the activity, discuss how they worked out which combinations of fractions, decimals and percentages totalled one. Also discuss which was their preferred representation and whether or not they converted all the values into like terms to find the solution to the puzzle.

Let's Investigate

- Encourage the children to start with simple fractions.
- Most children should be able to find rules for one and two decimal places.
- Some children will realise that fractions with denominators of 3, 6, 7, 9, 11, … have recurring digits, but are unable to identify the rule. If so, encourage them to look for the link between these denominators.

AfL

- What patterns / relationships do you notice? What conclusions can you draw?
- What is the equivalent fraction / decimal / percentage?
- How did you work out the equivalent fraction / decimal / percentage?
- How did you solve this puzzle? How did you begin?
- What conclusions / generalisations can you make?

Answers

Page 1

The Puzzler

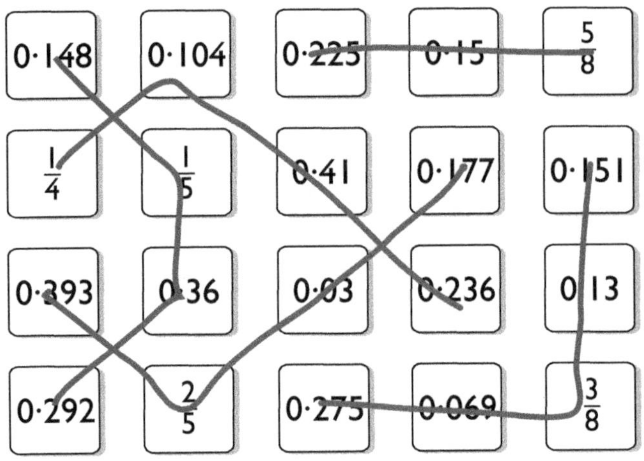

The Language of Maths

The three dots indicate that the decimal places continue, i.e.
$\frac{1}{7} = 0\cdot142857142857\ldots$

The two dots indicate that the digits below them and between them keep repeating in the same order, i.e.

$\frac{2}{7} = 0\cdot\dot{2}8571\dot{4}$
$\frac{3}{7} = 0\cdot\dot{4}2857\dot{1}$
$\frac{4}{7} = 0\cdot\dot{5}7142\dot{8}$
$\frac{5}{7} = 0\cdot\dot{7}1428\dot{5}$
$\frac{6}{7} = 0\cdot\dot{8}5714\dot{2}$
$\frac{1}{13} = 0\cdot\dot{0}7692\dot{3}$
$\frac{2}{13} = 0\cdot\dot{1}5384\dot{6}$
$\frac{3}{13} = 0\cdot\dot{2}3076\dot{9}$
$\frac{4}{13} = 0\cdot\dot{3}0769\dot{2}$
$\frac{5}{13} = 0\cdot\dot{3}8461\dot{5}$
$\frac{6}{13} = 0\cdot\dot{4}6153\dot{8}$
$\frac{7}{13} = 0\cdot\dot{5}3846\dot{1}$
$\frac{8}{13} = 0\cdot\dot{6}1538\dot{4}$
$\frac{9}{13} = 0\cdot\dot{6}9230\dot{7}$
$\frac{10}{13} = 0\cdot\dot{7}6923\dot{0}$
$\frac{11}{13} = 0\cdot\dot{8}4615\dot{3}$
$\frac{12}{13} = 0\cdot\dot{9}2307\dot{6}$

Page 2

Looking for Patterns

$\frac{2}{7}$ and $\frac{7}{11}$ are recurring decimals.
Recurring decimals make regular patterns.
Results of the investigation will vary.

Page 3

The Puzzler

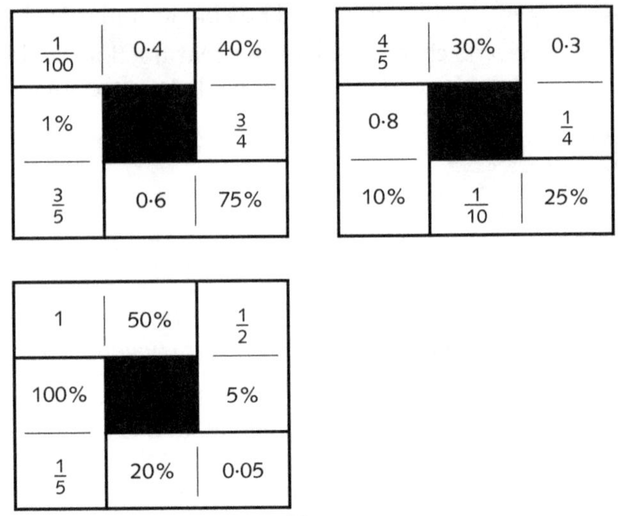

Other arrangements are possible.

Looking for Patterns

Results of the investigation will vary.
The shaded and the unshaded areas of each 100 square together, when they are expressed as percentages, always total 100%.

Page 4

The Puzzler

$\frac{2}{5}$	+	0·42	+	18%	= 1
$\frac{9}{25}$	+	0·21	+	43%	= 1
$\frac{6}{25}$	+	0·73	+	3%	= 1
$\frac{3}{10}$	+	0·55	+	15%	= 1
$\frac{23}{100}$	+	0·6	+	17%	= 1
$\frac{7}{50}$	+	0·38	+	48%	= 1
$\frac{7}{20}$	+	0·12	+	53%	= 1
$\frac{1}{20}$	+	0·28	+	67%	= 1

Let's Investigate

Rules are:

- One decimal place: fractions with a one-digit numerator and 10 as the denominator (or their equivalent fractions), i.e. $\frac{1}{10}, \frac{1}{5}, \frac{3}{10}, \frac{2}{5}, \frac{1}{2}, \ldots$
- Two decimal places: fractions with one- or two-digit numerators and 100 as the denominator (or their equivalent fractions) that are not equivalent to the group above, i.e. $\frac{1}{100}, \frac{1}{50}, \frac{3}{100}, \frac{1}{25}, \frac{1}{20}, \frac{3}{50} \ldots \frac{11}{100}, \frac{3}{25}, \frac{13}{100}, \ldots$
- Three decimal places: fractions with one-, two- or three-digit numerators and 1000 as the denominator (or their equivalent fractions) that are not equivalent to either of the above groups i.e. $\frac{1}{1000}, \frac{1}{500}, \frac{3}{1000}, \frac{1}{250}, \frac{1}{200}, \frac{3}{500} \ldots \frac{101}{1000}, \ldots$
- Recurring decimal: any fraction which, when expressed in its simplest form, has a denominator which is a multiple of a prime number other than 2 or 5, i.e. 3, 6, 7, 9, 11, 12, 13, 14, …

Inquisitive ant

Proportion
A part to whole comparison.

Issue 19

Length

Prerequisites for learning

- Read, choose, use and record standard metric units to estimate and measure length to a suitable degree of accuracy
- Begin to understand compound units for speed, such as kilometres per hour
- Draw and measure lines to the nearest millimetre
- Calculate mentally with integers and decimals
- Use formal written methods to add, subtract, multiply and divide integers and decimals

Resources

pencil and paper
Resource sheet 2: My notes (optional)
Resource sheet 3: Pupil self assessment booklet (optional)
Resource sheet 19: 2 cm squared paper
ruler
map of the local area
map of the world
tape measure
trundle wheel
calculator
sports cones
computer with Internet access

Teaching support

Page 1

Around the World

- Use as large a scale of map as possible.
- Explain to the children the difference, if necessary, between 'as the crow flies' and actual distances using roads, paths etc. Perhaps ask children to estimate actual distances and compare them to distances as the crow flies.

Let's Investigate

- Ask children to mark on the grid different pairs of points that are 6 units / 9 units / 8·5 units … apart.

What's the Problem?

- 1 million metres (1000 km) is approximately the distance from London to Monaco, as the crow flies.
- 1 million kilometres is approximately 2·5 times the distance of the Earth from the Moon, or 25 times the circumference of the Earth.

Page 2

Let's Investigate

- Children need to put a known number of layers of paper or clothing together to make a thickness they can measure.

Around the World

- Ask children to estimate the distance that Phileas Fogg travelled.

What's the Problem?

- Number of hours travelled: 455 ÷ 65 = 7 hours.
- 7 hours prior to 6:25 pm is 11:25 am.

Page 3

What's the Problem?

- Assist the children in drawing a diagram, i.e.

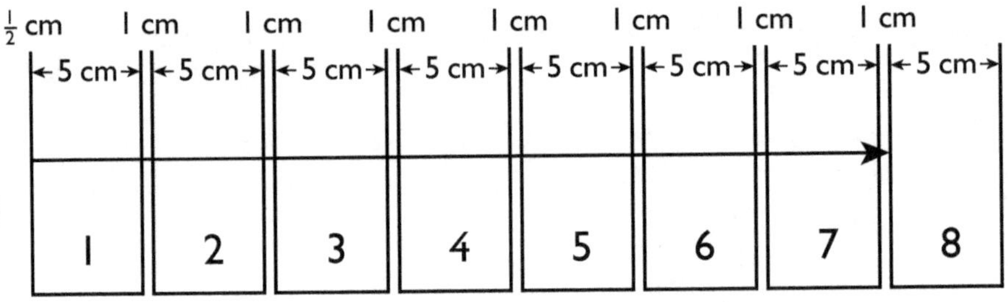

Therefore the total number of centimetres that the moth ate is:

$\frac{1}{2}$ cm + (7 × 1 cm) + (7 × 5 cm)

$\frac{1}{2}$ cm + 7 cm + 35 cm

$42\frac{1}{2}$ cm

What's the Problem?

- Suggest children use trial and improvement to work out the answer. They need to estimate two speeds that have a difference of 5 km/h. They then multiply the slower speed by eight hours, and the faster speed by half an hour less, i.e. $7\frac{1}{2}$ hours. If the two products are the same, then their estimate is correct. If the two products differ, then they need to adjust their speeds accordingly.

Focus on Science

- Each day the caterpillar climbs up 10 cm and then slips back 6 cm overnight; so in a 24-hour period it gets 4 cm higher up the trellis.
- After 48 days / nights the caterpillar has climbed / fallen to 192 cm. On the 49th day it need only climb eight cm to reach the top of the trellis.

Page 4

What's the Problem?

• Paul:	60 km	140 km	220 km	300 km	380 km	460 km	540 km
		|	|	|	|	|	|
		1 h	2 h	3 h	4 h	5 h	6 h
		|	|	|	|	|	|
Derek:		90 km	180 km	270 km	360 km	450 km	540 km

Sports Update

- Remind children that the size of their step is the distance from the front of one foot to the front of the other foot after they have taken a step (not the distance between the front of one foot and the back of the other). If they are having difficulty measuring one step, suggest that they measure the distance of five or ten steps, and then divide by five or ten.

At Home

- Remind children of the effect of perspective when estimating size: that the further in front or behind them an object is, the larger or smaller than its actual size it will appear.
- Once the children have completed the activity, ensure that there is an opportunity in class for pairs or groups of children to discuss their results.

Construct

- Suggest the children lay out cones so that they form the vertices of a square (or other polygon) with sides of approximately 5 / 12 / 15 … metres in length.

AfL

- What is the scale of your map? What does this mean? What is the distance between these two points on the map?
- What is the shortest route? How do you know that it's the shortest route?
- What do your estimates / approximations mean? How accurate are they? What does this mean?
- What were the results of your investigation? What conclusions can you draw?
- How did you work out the answer to this problem / puzzle? What did you use to help you?

Answers

Page 1

Around the World
Answers will vary.

Let's Investigate
Answers will vary.

What's the Problem?
1 million metres are 1000 km.
Answers will vary.

Page 2

Let's Investigate
Results of the investigation will vary.

Around the World
Phileas Fogg travelled by train, steam boat, elephant and wind-powered sledge.
The route went from London to Suez, Bombay, Calcutta, Hong Kong, Shanghai, Yokohama, San Francisco, New York, Dublin and London.

What's the Problem?
11:25 am

Page 3

What's the Problem?
$42\frac{1}{2}$ cm

What's the Problem?
75 km per hour
80 km per hour

Focus on Science
49 days

Page 4

What's the Problem?
6 hours

Sports Update
Answers will vary.

At Home
Answers will vary.

Construct
Answers will vary.

Inquisitive ant

tacheometer (or tachymeter)
An instrument used in surveying to work out distances and heights.

Issue 20

Mass

Prerequisites for learning

- Read, choose, use and record standard metric units to estimate and measure mass to a suitable degree of accuracy
- Understand and use approximate equivalences between metric units and common imperial units
- Calculate mentally with integers and decimals
- Use formal written methods to add, subtract, multiply and divide integers and decimals
- Answer a question by collecting, selecting and organising relevant data; draw conclusions, using ICT to present features

Resources

pencil and paper
Resource sheet 2: My notes (optional)
Resource sheet 3: Pupil self assessment booklet (optional)
calculator
ruler
weighing equipment
collection of different carrier bags
data handling software
newspapers, magazines, food packaging, telephone directory, textbook
selection of items suitable for sending through the post, e.g. book, board game, pair of shoes in a shoe box, packet of pencils or felt tips
computer with Internet access

Teaching support

Page 1

Focus on Science

- Plastic supermarket carrier bags are quite strong so, if putting weights into them to test their strength, suggest that they are suspended close to the ground.

What's the Problem?

- One method of working out the answer is, through trial and improvement, to assign values to each of the fruits. For example, orange = 5, lemon = 3 and clementine = 1.

  three oranges and two lemons weigh the same as four oranges and one clementine.

  $(3 \times 5) + (2 \times 3) = (4 \times 5) + (1 \times 1)$

  $15 + 6 = 20 + 1$

  $21 = 21$

  One orange weighs the same as one lemon and two clementines.

  $(1 \times 5) = (1 \times 3) + (2 \times 1)$

  $5 = 3 + 2$

  $5 = 5$

  How many clementines weigh the same as one orange?

  $(5 \times 1) = 5$

  $5 = 5$

- Suggest the following strategy to the children:

  Draw a large scale balance, balanced equally. On one side draw three oranges and two lemons, on the other side four oranges and one clementine. (three oranges and two lemons have the same mass as four oranges and one clementine.)

  Now suggest that they take away one orange from the left-hand side (cross it out), and in its place draw one lemon and two clementines (one orange has the same mass as one lemon and two clementines). Keep doing this until all the oranges have been replaced by lemons and clementines.

  There are now five lemons and six clementines on the left-hand side and four lemons and nine clementines on the right.

Remind children that if they take away a lemon from each side, the scale will remain balanced. Do this, one at a time from each side (cross them out), until there is one lemon left on the left-hand side. Now do the same with the clementines until there are three left on the right-hand side. one lemon = three clementines.

Repeat the procedure for the statement 'one orange has the same mass as one lemon and two clementines', replacing the lemon with three clementines, to get one orange = five clementines.

What's the Problem?

- For every kilogram picked, 900 g are sold (1000 g – 100 g)
  Every day 145 kg of strawberries are picked.
  145 × 900 g = 130 500 g (130·5 kg)
  So in one day The William's Strawberry Farm is able to sell 130·5 kilograms of strawberries.
  There are a total of 61 days in May and June.
  130·5 × 61 = 7960·5 kilograms of strawberries are sold in May and June.

Page 2

What's the Problem?

- Discuss with the children trial and improvement strategies for deciding which two or more weights are likely to total a larger amount, e.g. by referring to the most significant digits to determine the magnitude and / or by referring to the least significant digits in the weights to determine the least significant digit in the larger amount.

Let's Investigate

- For this activity children will need access to a stack of the same paper as this Issue is printed on.
- Remind children that the size, or area, of the paper and cardboard they are measuring is important when comparing the different weights, i.e. weight per square centimetre or per square metre.

In the Past

- Children investigate weighing systems from other civilisations, e.g. Indus Valley Civilization.

Page 3

In the Past

- Ask the children to offer suggestions as to why they think many of the imperial units of weight are no longer in common usage.

In the Past

- This activity links with the Inquisitive ant feature for this issue, i.e. Avoirdupois weight.

Page 4

Let's Investigate

- One adult portion is 80 g of fruit or vegetables. Recommended portion size for children varies depending on a child's age and body size, but is roughly the amount that will fit in the palm of one hand.

Let's Investigate

		1st Class	2nd Class
Size	**Weight up to and including**	**Price**	**Price**
Small Parcel	1 kg	£3·30	£2·80
Not exceeding: length 45 cm, width 35 cm, depth 16 cm.	2 kg	£5·45	£2·80
Medium Parcel	1 kg	£5·65	£4·89
Not exceeding: length 61 cm, width 46 cm, depth 46 cm.	2 kg	£8·90	£4·89
	5 kg	£15·85	£13·75
	10 kg	£21·90	£20·25
	20 kg	£33·40	£28·55

NOTE: Larger parcels are only available on Parcelforce Worldwide services.

- The cost of sending an item abroad does not only depend upon its measurements and weight, but also the parcel's destination.
- When sending items abroad, children will also need to decide whether to send them airmail or surface mail.
- As there are many variables when sending an item abroad, there are many different pricing levels. Therefore children will need to consult the Royal Mail website.

AfL

- What did you find out? How did you record your results? Why did you decide to present them in this way?
- What conclusions can you draw from your investigation / experiment?
- How did you work out the problem / puzzle? What strategies did you use? What calculations did you do?
- What relationships can you see in the different measures?
- How do the words used for some of the measures convey their meaning?

Answers

Page 1

Focus on Science
Results of the investigation will vary.

What's the Problem?
5 clementines have the same mass as 1 orange.

What's the Problem?
7960·5 kilograms

Page 2

What's the Problem?
638·25 kg = 263·45 kg + 374·8 kg
821·927 kg = 263·45 kg + 406·137 kg + 152·34 kg
1087·418 kg = 406·137 kg + 374·8 kg + 306·481 kg

Let's Investigate
Answers will vary.

In the Past
The Babylonians and Sumerians had a system in which there were 60 *shekels* in a *mina*, 60 *minas* in a *talent* and 3600 *shekels* in a *talent*.

The Roman *talent* consisted of 100 *libra* (pound) which were smaller in size than the *mina*.

Page 3

In the Past
16 drams = 1 ounce
16 ounces = 1 pound
7 pounds = 1 clove
14 pounds = 1 stone
28 pounds = 1 tod
112 pounds = 1 hundredweight
364 pounds = 1 sack
2240 pounds = 1 ton
2 stones = 1 quarter
4 quarters = 1 hundredweight
20 hundredweight = 1 ton

In the Past
The basic unit of weight in the British system is the grain – originally based on the weight of a grain of barley.
The Avoirdupois pound, used today, is made up of 7000 grains, and is split into 16 ounces (each ounce is therefore 437·5 grains).

Page 4

Let's Investigate
Answers will vary.

Let's Investigate
Results of the investigation will vary.

Inquisitive ant

Avoirdupois weight
A system for measuring weights based on the pound.

Issue 21

Volume and capacity

Prerequisites for learning

- Read, choose, use and record standard metric units to estimate and measure volume/capacity to a suitable degree of accuracy
- Understand and use approximate equivalences between metric units and common imperial units
- Calculate mentally with integers and decimals
- Use formal written methods to add, subtract, multiply and divide integers and decimals

Resources

pencil and paper
Resource sheet 2: My notes (optional)
Resource sheet 3: Pupil self assessment booklet (optional)
Resource sheet 18: 1 cm squared paper
Resource sheet 19: 2 cm squared paper
ruler
calculator
scissors
1 litre bucket of water
weighing equipment
computer with Internet access

Teaching support

Page 1

At Home

- If children do not have a dishwasher at home, you may wish to discuss with them the amount of water that is used when washing dishes by hand. Hand washing dishes can be efficient if you use a bowl and watch how much water you use. However, daily hand washing of dishes typically uses about 63 litres. If dishes are rinsed off under a running tap the total water used can increase to 150 litres.
- Children should count the number of times a day (or week) that their family does each of the actions on the list and add up the total amount of water used, using the figures given.
- Once the children have completed the activity, ensure that there is an opportunity in class for pairs or groups of children to discuss their results.

What's the Problem?

- Explain how to measure one litre of water with the children first, so that they see the method (see Answers).

The Puzzler

- Discuss with the children trial and improvement strategies for deciding which three or more capacities are likely to give one of the larger amounts.

Page 2

Construct

- Ensure children understand what they are required to do for this activity.
- Remind children that volume is the space actually occupied by an object or the bulk of some substance. In the case of the activity, it refers to the size of the three-dimensional space enclosed within the sides of each of the boxes.
- Also remind the children of the formula for calculating the volume of a cuboid:
  length × width × height
  or $l \times w \times h$
  or lwh

At Home

- Once the children have completed the activity, ensure that there is an opportunity in class for pairs or groups of children to discuss their results.
- Children investigate the cost of petrol and diesel in different countries in Europe. Which is more expensive in each of these countries, petrol or diesel?

Page 3

In the Past

- Children investigate other systems for measuring volume and capacity from other civilisations, e.g. Ancient Egyptian.

Let's Investigate

- Many children will probably already know that one litre of water has a mass of one kilogram. However, the important aspect of this activity is for them to design an experiment to prove it.

In the Past

- Children investigate other systems for measuring volume and capacity from other civilisations, e.g. Ancient Egyptian.

Page 4

In the Past

- Children investigate other historical measurements for volume and capacity in England, e.g.
  Mouthful (about $\frac{1}{2}$ ounce)
  Pony (2 mouthfuls or 1 ounce)
  Jigger (1·5 ounces)
  Jack or Jackpot (2 ounces)
  Pottle or Half Gallon (80 oz)
  Kenning or Pail (2 pecks or 4 gallons)
  Strike (2 bushels or 16 gallons)
  Coomb (2 strikes or 32 gallons)
  Cask (2 coombs or 64 gallons)

Around the World

- Children investigate the difference between the UK ton ('long ton' – 2240 pounds) and the US ton ('short ton' – 2000 pounds).

AfL

- How do you calculate volume? What about for a different solid?
- What calculations did you perform to work out that answer / approximation?
- What can you tell me about how volume and capacity were measured by the Babylonians / Ancient Greeks?
- What were the results of your experiment? What conclusions can you make?
- What did your investigation discover? Why do you think this is?
- What relationships can you see in the different measures? What are the similarities / differences?
- How do the words used for some of the measures convey their meaning?

Answers

Page 1

At Home
Answers will vary.

What's the Problem?
1 litre = Fill the 5-litre and the 3-litre buckets from the 9-litre bucket. 1-litre remains in the 9-litre bucket.
2-litres = Fill the 5-litre bucket. From this bucket fill the 3-litre bucket. 2-litres remain in the 5-litre bucket.
3-litres = Fill the 3-litre bucket.
4-litres = Fill the 5-litre bucket from the 9-litre bucket. 4 litres remain in the 9-litre bucket.
5-litres = Fill the 5-litre bucket.
6-litres = Fill the 3-litre bucket from the 9-litre bucket. 6-litres remain in the 9-litre bucket.
7-litres = Fill the 5-litre bucket from the 9-litre bucket. 4-litres remain in the 9-litre bucket. Then, from the 5-litre bucket fill the 3-litre bucket. Add the 3-litres to the 4-litres remaining in the 9-litre bucket to make 7-litres.
8-litres = Fill the 3-litre and 5-litre buckets from the 9-litre bucket.
9-litres = Fill the 9-litre bucket.

The Puzzler
a. 475 litres + 1538 litres – 1837 litres = 176 litres
b. 1538 litres – 846 litres + 1837 litres = 2529 litres
c. 2338 litres – 846 litres + 1837 litres – 475 litres = 2854 litres
d. 2338 litres – 475 litres – 1538 litres + 846 litres = 1171 litres
e. 846 litres + 1538 litres – 1837 litres + 2338 litres = 2885 litres

Page 2

Construct
Volume of the first box is 1 cm × 8 cm × 8 cm = 64 cm^3
Volume of the second box is 2 cm × 6 cm × 6 cm = 72 cm^3
Volume of the third box is 3 cm × 4 cm × 4 cm = 48 cm^3
Volume of fourth box is 4 cm × 2 cm × 2 cm = 16 cm^3
So, the fourth box has the smallest volume.
So, the second box has the greatest volume.

At Home
Answers will vary.
Diesel is more expensive than petrol.

Page 3

In the Past
One *sila* is approximately equal to one litre.
One *sila* was equal to 60 *gin*. One *gin* was equal to 180 *she*.

Let's Investigate
True. 1 litre of water weighs 1 kg. 500 ml of water weighs 500 g.

In the Past
The *cotyle* or *cotyla* ranged from 210 ml to 330 ml (about the capacity of a cup or a can of drink).
The *choenix* was about the volume of a man's daily ration of grain.
A *wine amphora* held 39 litres of wine.

Page 4

In the Past

60 minims	1 fluid drachm
8 fluid drachms	1 fluid ounce
20 fluid ounces	1 pint
4 gills	1 pint
2 pints	1 quart
4 quarts	1 gallon
2 gallons	1 peck

4 pecks	1 bushel
8 bushels	1 quarter
36 bushels	1 chaldron
9 gallons	1 firkin
4 firkins	1 barrel
$52\frac{1}{2}$ gallons	1 hogshead
$26\frac{2}{3}$ fluid ounces	1 bottle

Around the World

	gallon	pint	fluid ounce
UK	8 UK pints or ≈ 4·55 litres	0·568 litre	$\frac{1}{20}$ UK pint or ≈ 28·41 ml
US	8 US pints or ≈ 3·79 litres	0·473 litre	$\frac{1}{16}$ US pint or ≈ 29·57 ml

Inquisitive ant

hectolitre
Equal to 100 litres.

Issue 22

Time

Prerequisites for learning

- Solve problems involving converting between units of time
- Use simple expressions and formulae in symbols
- Calculate mentally with integers and decimals
- Use formal written methods to add, subtract, multiply and divide integers and decimals

Resources

pencil and paper
Resource sheet 2: My notes (optional)
Resource sheet 3: Pupil self assessment booklet (optional)
analogue clock with geared hands (optional)
calculator
map of the world showing time zones
coloured pencils
computer with Internet access

Teaching support

Page 1

Around the World

- Ensure children are familiar with the terms *longitude, latitude* and *Greenwich Mean Time* (GMT).
- Referring to the time zone map, you may wish to discuss with the children that numbers written as +1, +2, +3 … refer to the number of hours ahead of GMT, and that numbers written as –1, –2, –3 … refer to the number of hours behind GMT.

Around the World

- Ensure the children are familiar with a 24-hour timeline.
- This activity relies on children having completed the Around the World activity above, and realising the time difference between their current time zone and the current time zone in Sydney, Australia.

Page 2

Famous Mathematicians

- Work through one child's birthday so that the children understand how to use the tables.

Page 3

Let's Investigate

- This activity relies on children having completed the Famous Mathematicians activity on page 2.
- When the children have completed this activity and the Famous Mathematicians activity on page 2, discuss with them which method (and why) they preferred for working out on which day of the week a person was born.

Page 4

Let's Investigate

- You may wish to ask the children to work out the first ten DDMMYYYY palindromic dates in the 21st century (see Answers).

What's the Problem?

- One method of working out this investigation is to time how long it takes to count 100 or 200 numbers and multiply up. Discuss with the children which numbers they should use for this method and why, e.g. the numbers 145 601 to 145 700 as opposed to 1 to 100.

At Home

- Children can complete this activity just as effectively in school as at home.
- If the children do undertake this activity at home, once they have completed the task, ensure that there is an opportunity in class for pairs or groups of children to discuss their results.

What's the Problem?

- Assuming that Lisa's great-grandma is 80 years old she could be 30 000 days old.
  365 × 80 = 29 200 days
  (365 days a year) × (80 years old)
- Assuming that Lisa's great-grandma sleeps eight hours a day on average and that she is 80 years old, then she could have slept for over 200 000 hours during her lifetime.
  8 × 365 × 80 = 233 600 hours
  (8 hours of sleep a day) × (365 days a year) × (80 years old).

AfL

- Can you explain to me how the world's time zones work?
- How did you work out the answer to the problem / puzzle? Talk me through your calculation. How did you work that out?
- Explain to me the formula for finding out on which day of the week a person was born.
- How does this method for finding out on which day of the week a person was born differ from the formula devised by Carl Gauss? Which method do you think is best? Why?

Answers

Page 1

Around the World
Answers will vary.

Around the World
Answers will vary.

Page 2

Famous Mathematicians
Answers will vary.

Page 3

Let's Investigate
Answers will vary.

Page 4

Let's Investigate
There are 29 in total, the first 10 are:
10 February 2001 (10022001)
20 February 2002 (20022002)
1 February 2010 (01022010)
11 February 2011 (11022011)
21 February 2012 (21022012)
2 February 2020 (02022020)
12 February 2021 (12022021)
22 February 2022 (22022022)
3 February 2030 (03022030)
13 February 2031 (13022031)

What's the Problem?
Answers will vary.

At Home
Answers will vary.

What's the Problem?
Lisa's great-grandma could almost be 30 000 days old if she is approximately 80 years old. She would have slept for more than 200 000 hours.
Explanations and other 'amazing facts' will vary.

Inquisitive ant

CET
Central European Time. CET is 1 hour ahead of Greenwich Mean Time (GMT).

Issue 23

Temperature

Prerequisites for learning

- Interpret and compare diagrams that represent data
- Construct frequency tables and line graphs to represent changes over time, using ICT where appropriate

Resources

pencil and paper
Resource sheet 2: My notes (optional)
Resource sheet 3: Pupil self assessment booklet (optional)
Resource sheet 18: 1 cm squared paper
ruler
thermometer
weather sections from different newspapers
computer with Internet access
data handling software

Teaching support

Page 1

Focus on Science

- Ensure children are familiar with the terms *longitude* and *latitude*.
- Ask children to investigate how the Humboldt current affects the west coast of the Americas.

Focus on Science

- Choosing one place inside the school buildings, one place outside and one place that has a large expanse of glass windows should give the greatest variety of temperatures.

Page 2

Around the World

- Ensure children write about the three different aspects of the activity, i.e.
  - the temperature differences at different times throughout the year in Paris
  - the temperature differences at different times throughout the year in Nice
  - the temperature differences between Paris and Nice at different times throughout the year.
- Suggest children comment on the seasonal differences between Paris and Nice.

Around the World

- Children may need help deciding on the range of numbers for the temperature axis of the graph.

Technology Today

- Children will need to be familiar with the line graph function in a data handling software package in order to complete this activity.
- If the software package being used does not facilitate the drawing of two lines on the same line graph, then suggest the children draw two separate line graphs.

Page 3

At Home

- Most recipe books include a temperature guide for both degrees Celsius and degrees Fahrenheit.
- You may wish the children to undertake this activity in school rather than at home. If so, suggest they use the Internet to find the information they require.
- If the children do undertake this activity at home, once they have completed the task ensure that there is an opportunity in class for pairs or groups of children to discuss their results.

Focus on Science

- Ensure children understand how the table works.
- To highlight the effect of wind-chill, discuss with the children what happens when you are in front of a fan.

Page 4

Focus on Science

- In order to use the Wind-chill Factor Chart on page 3, children will need to look for temperatures in the weather section of the newspapers that are in the range 10°C to $^-4$°C.
- Discuss with the children what the headings of the table should be. For example:

Location	Temperature	Wind speed	Temperature feels like

- If the weather sections from different newspapers do not include data about temperature and wind speed then the children will need to use the Internet to find out this information.

Focus on Science

- Children need to understand that, when drawing a graph with such a narrow temperature range, it is important not to start the scale with zero degrees and to have a scale large enough to show clearly the differences in temperature.

AfL

- What did you find out? How did you record your results? Why did you decide to present them in this way?
- What conclusions can you draw from your investigation / experiment?
- What things did you need to think about when drawing your line graph? What was similar / different when you used the data handling software?
- How do you convert this temperature of … into …?
- Explain to me what 'wind-chill factor' means. How is it calculated?

Answers

Page 1

Focus on Science
Answers will vary.

Focus on Science
Answers will vary.

Page 2

Around the World
Statements will vary.
Temperature difference in Paris between winter and summer is 18°C. Temperature difference in Nice between winter and summer is 16°C.
Temperature differences between Paris and Nice are greater in winter than in summer.

Around the World

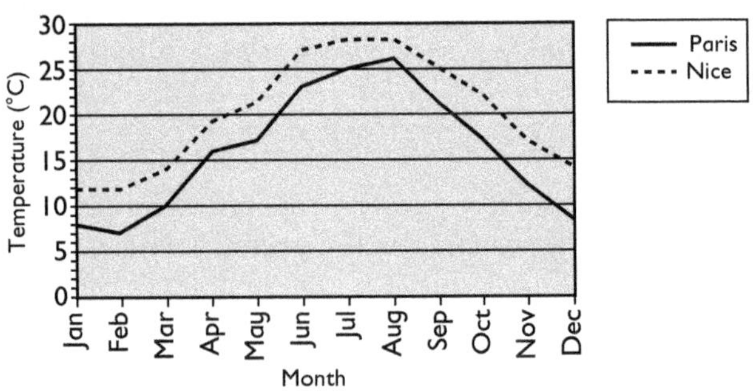

Technology Today
As above.

Page 3

At Home

	Temperature range in degrees Celsius	Temperature range in degrees Fahrenheit
Low	140–150	280–300
Moderate	160–180	320–360
Moderately hot	190–210	375–410
Hot	220–230	430–450
Very hot	240	460

Focus on Science
Answers will vary.

Page 4

Focus on Science
Answers will vary.

Focus on Science

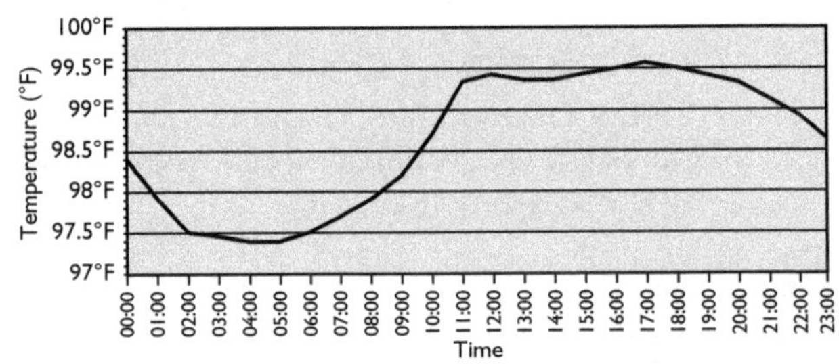

Inquisitive ant

forecast
A prediction of the weather conditions for the near future. Forecasts are usually broadcast on television and radio and printed in newspapers. You can also find them on the Internet.

Issue 24

Area

Prerequisites for learning

- Calculate and compare the areas of rectangles and estimate the area of irregular shapes
- Calculate the area from scale drawings using given dimensions
- Calculate surface area of cubes and cuboids
- Understand and use simple algebraic conventions
- Use simple expressions and formulae in words and symbols
- Calculate mentally with integers
- Use formal written methods to add, subtract, multiply and divide integers

Resources

pencil and paper
Resource sheet 2: My notes (optional)
Resource sheet 3: Pupil self assessment booklet (optional)
Resource sheet 18: 1 cm squared paper
Resource sheet 21: Squared dot paper (optional)
scissors
ruler
calculator
matchsticks
Peters Projection map of the world
computer with Internet access

Teaching support

Page 1

The Puzzler

- Suggest children draw the kite, cut it in half vertically and then cut one of the halves in two along the axis. Now tell them to rearrange the pieces to make a rectangle.

The Puzzler

- Suggest children draw the shapes on squared dot paper to better show the matchsticks, e.g.

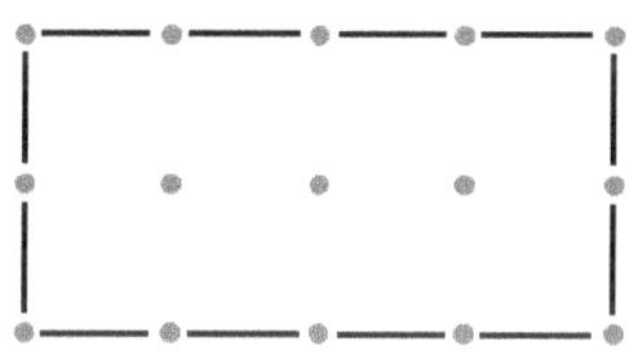

Page 2

Around the World

- If you do not have access to a Peters Projection map of the world, then the children will need to locate, download and print one from the Internet.
- Where the outlines of the countries cut across squares on the paper, children need to estimate how much of the square is occupied by the area of the country.

Focus on Science

- Where the outlines of hands and feet cut across squares on the paper, children need to estimate how much of the square is occupied by the area of the hand or foot.

Construct

- Unlike the previous two activities, the areas of the rectangles can be calculated exactly without requiring estimation.
- Point out one pair of squares with an outline cutting across them which together make a whole square.

Page 3

What's the Problem?

- Children can approach this problem by making an estimate of the measurements of one face (preferably the smallest face as there will be fewer possibilities) and, using this estimate, work out the dimensions of the other two faces. If the estimate does not fit, they refine it through trial and improvement until they find the correct dimensions.

What's the Problem?

- If necessary, discuss with the children the algebraic convention for omitting the multiplication sign when writing an algebraic expression, i.e. 2(3 m × 2·5 m) + 2(2 m × 2·5 m) is the same as 2 × (3 m × 2·5 m) + 2 × (2 m × 2·5 m).

Page 4

What's the Problem?

- Area of square EFGH is 18 cm^2 (36 cm^2 ÷ 2).
  Area of square IJKL is 9 cm^2 (18 cm^2 ÷ 2).
  Area of square MNOP is 4·5 cm^2 (9 cm^2 ÷ 2).
  The area of the shaded space is 13·5 cm^2.
  (18 cm^2 – 4·5 cm^2 is the difference between MNOP and EFGH.)

The Puzzler

- FB is half the length of BC so BC = 12 cm.
  Area of BCEF = 12 cm × 6 cm = 72 cm^2.
  Area of CEF = $\frac{1}{2}$ area of BCEF = 36 cm^2.
  Area of ABF = area of CDE = 6 cm × 3 cm = 18 cm^2.
  So total area of shape = 2 × 18 cm^2 + 72 cm^2 = 108 cm^2.

Let's Investigate

- For this activity, children need to find out the coverage for both paint and wallpaper.
  - One litre of paint will cover approximately 12 m^2 per coat.
  - Most wallpaper rolls are made to a standard size of 10·5 m long and about 53 cm wide. The table below shows the number of rolls needed for various room sizes.

		Distance round room (including doors and windows)							
		10 m	12 m	14 m	16 m	18 m	20 m	22 m	24 m
Wall height (*h*)	2·1 m < $h \le$ 2·3 m	5	5	6	7	8	9	10	11
	2·3 m < $h \le$ 2·4 m	5	6	7	8	9	10	10	11
	2·4 m < $h \le$ 2·6 m	5	6	7	9	10	11	12	13
	2·6 m < $h \le$ 2·7 m	5	6	7	9	10	11	12	13
	2·7 m < $h \le$ 2·9 m	6	7	8	9	10	12	12	14

- Children should be made aware that they would probably need to give the walls two coats of paint, and that this needs to be taken into consideration when calculating the cost of painting their classroom. Likewise, when calculating the cost of wallpaper, they need to think about the cost of wallpaper paste (glue).
- Children should also think about the other costs involved in painting / wallpapering, e.g. paint / glue tray; brush / roller; dust sheets.

What's the Problem?

- If the area of square ABCD is 36 cm^2, then each side of the square is 6 cm. So if the lengths of sides AB and AD are 6 cm each, then the lengths of AE and AH are 3 cm each.
- Given that the area of a triangle is $\frac{1}{2}lh$, then the area of triangle AEH is $\frac{1}{2}$ × (3 cm × 3 cm) = 4·5 cm^2.

AfL

- How did you work out the answer to the problem / puzzle? Could you see any patterns / relationships?
- How does a Peters Projection map of the world vary from the conventional map of the world? Which is best? Why?
- Explain to me how you devised your experiment / test. What things did you consider to be important?
- How did you work out the areas of these shapes? What calculations did you perform?
- What solution / answer did you get for this puzzle / problem? How did you solve it? What patterns did you notice? How did these help?

Answers

Page 1

The Puzzler
12 cm^2

The Puzzler

7 matchsticks

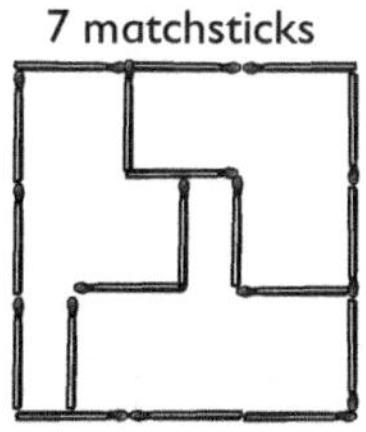

8 square units

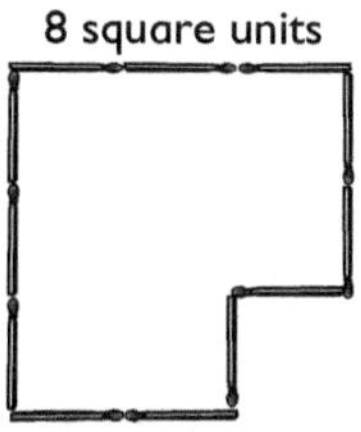

7 square units

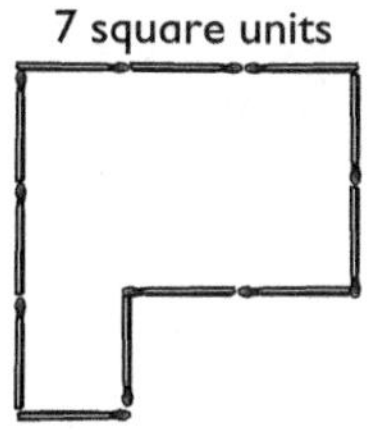

6 square units

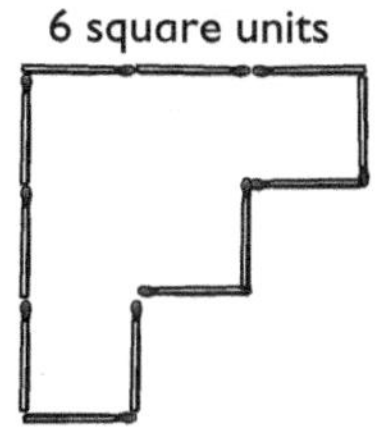

5 square units

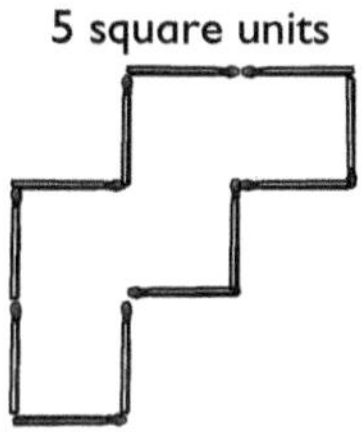

4 square units

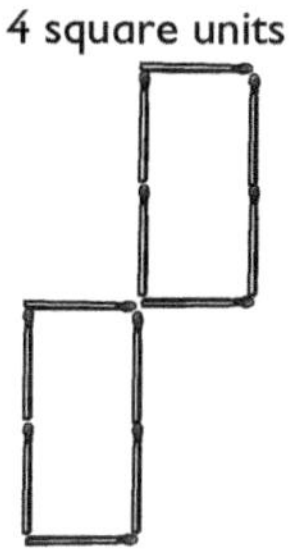

3 square units

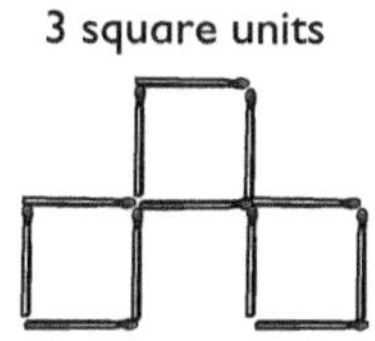

Other answers are possible.

Page 2

Around the World
Answers will vary.

Focus on Science
Answers will vary.

Construct
Answers will vary.

Page 3

What's the Problem?

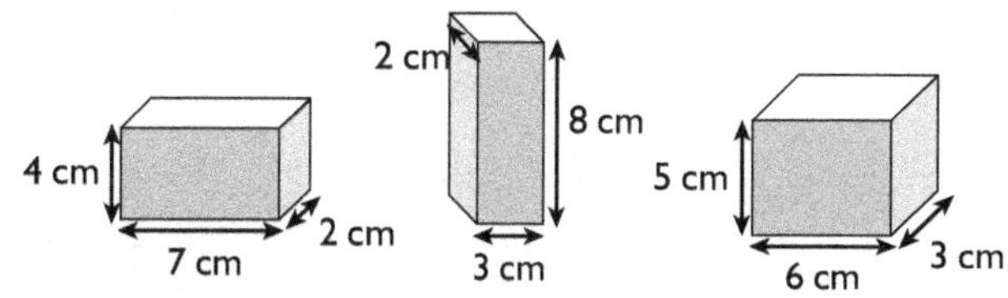

What's the Problem?
2(3 m × 2·5 m) + 2(2 m × 2·5 m) = 25 m^2
Richard will need to buy 625 tiles.
He will need to buy 25 boxes.
He will have to pay £312.50.

Page 4

What's the Problem?
The area of the shaded space is 13·5 cm^2.

The Puzzler
The area of rectangle BCEF is 72 cm^2.
The area of triangle CEF is 36 cm^2.
The total area of the shape is 108 cm^2.

Let's Investigate
Answers will vary.

What's the Problem?
The area of triangle AEH is 4·5 cm^2.

Inquisitive ant

acre
A unit of land area equal to 4840 square yards, 4046·86 square metres or 0·405 of a hectare.

Issue 25

Area and perimeter

Prerequisites for learning

- Calculate and compare the area and perimeter of rectilinear shapes
- Recognise that shapes with the same areas can have different perimeters and vice versa
- Calculate mentally with integers and decimals
- Use formal written methods to add, subtract, multiply and divide integers and decimals

Resources

pencil and paper
Resource sheet 2: My notes (optional)
Resource sheet 3: Pupil self assessment booklet (optional)
Resource sheet 18: 1 cm squared paper
Resource sheet 19: 2 cm squared paper
Resource sheet 21: Squared dot paper
Resource sheet 22: Triangular dot paper
ruler
0–9 dice
trundle wheel
stopwatch
interlocking cubes (or similar)
computer with Internet access

Teaching support

Page 1

What's the Problem?

- Given the dimensions of the pool and the border, the dimensions of the outside of the border are:

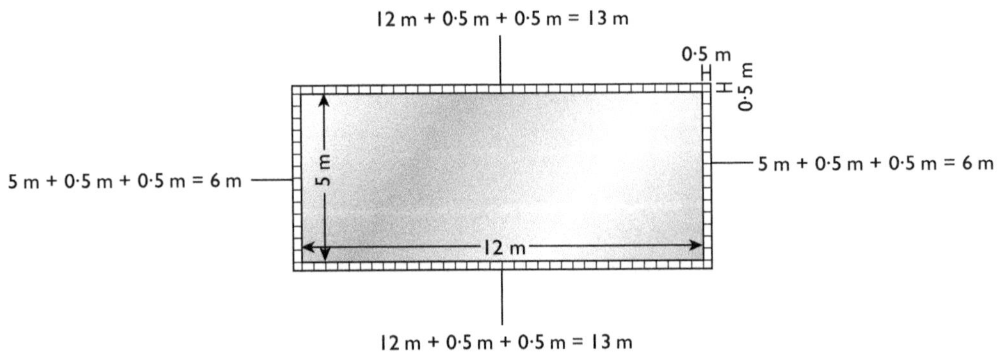

What's the Problem?

- It is recommended that children complete this activity before starting on the What's the Problem? activity at the top of page 2.
- Children can approach this problem by making an estimate of the dimensions of each rectangle and, using this estimate, work out its area and perimeter. If the estimate does not fit, they refine it through trial and improvement to find the correct dimensions.

Let's Investigate

- It is recommended that children complete this activity before starting on the second Let's Investigate activity at the top of page 3.
- The purpose of this activity is for children to investigate what different shapes are possible with the same perimeter.

Page 2

What's the Problem?

- It is recommended that children complete the second What's the Problem? activity on page 1 before starting this activity.
- It will help if children write the measurements of each allotment on the diagram as they work them out. Once they have worked out the sizes of A, D, E, F and J, they will have one measurement for each of B, C, H and I, and two measurements for G.

What's the Problem?

- It will help to write the sizes of all the allotments in the previous activity onto the diagram, to get the total size needed for this activity.

Page 3

Let's Investigate

- Ensure children know how to use, and gather measurements from, a trundle wheel.
- Probably the simplest way for children to work out how many circuits of the perimeter they would run in half an hour is to measure the time it takes to run one circuit and then divide 30 minutes by this time.

Let's Investigate

- It is recommended that children complete the Let's Investigate activity on page 1 before starting this activity.

Let's Investigate

- Children will need to use a variety of mathematical papers for this activity. The papers they require will depend upon the polygons they are investigating.

Page 4

Sports Update

- The size of a football pitch varies, but since 2007 the size for international football matches has been standardised at 105 m × 68 m.
- You may wish to introduce children to the formula for finding the circumference of a circle ($\pi \times$ diameter or πd). If so, they can calculate the lengths of the lines of the centre circle and quarter circles. They will still need to estimate the length of the penalty arcs.

Let's Investigate

- You may wish to suggest to the children that they only investigate perimeters of regular polygons. If so, children should then be able to arrive at the formula $n \times a$, where n is the number of sides and a is the length of one of the sides.

The Puzzler

- Suggest children use interlocking cubes (or similar) to help them work out the solutions to the puzzles.

AfL

- How did you work out the answer to this problem / puzzle? Could you have used an alternative method?
- What relationships do you notice?
- What alternative answers did you come up with?
- What are the results of your investigation?
- Are the results what you expected? How are they different? Why do you think this?

Answers

Page 1

What's the Problem?
Perimeter is 38 m.
Area is 18 m².

What's the Problem?
A: 4 cm × 12 cm
B: 7 cm × 10 cm
C: 6 cm × 15 cm
D: 4 cm × 24 cm
E: 12 cm × 12 cm

Let's Investigate
Answers will vary.

Page 2

What's the Problem?
A: 16 m × 9 m
B: 9 m × 9 m
C: 25 m × 9 m
D: 9 m × 15 m
E: 12 m × 15 m
F: 12 m × 9 m
G: (6 m × 9 m) + (9 m × 3 m)
H: 21 m × 6 m
I: 18 m × 12 m
J: 9 m × 18 m

What's the Problem?
Answers will vary.

Page 3

Let's Investigate
Answers will vary.

Let's Investigate
Answers will vary.

Let's Investigate
The area of shape 'B' is four times that of shape 'A'.
The area of shape 'C' is nine times that of shape 'A'.
The perimeter of shape 'B' is twice that of shape 'A'.
The perimeter of shape 'C' is three times that of shape 'A'.

Page 4

Sports Update
Answers will vary depending on the pitch size used.
The perimeter of the football pitch is 346 m.
There are 643·4 m of lines.

Let's Investigate
Answers will vary.

The Puzzler
If one of the corner squares is removed the perimeter will stay the same.
Other solutions are possible.

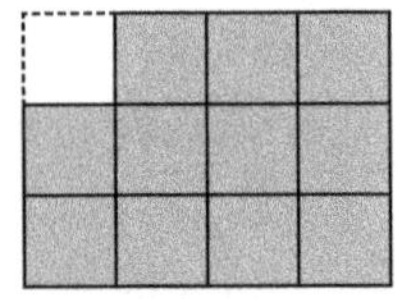

Eight squares can be removed without changing the perimeter.
Other solutions are possible.

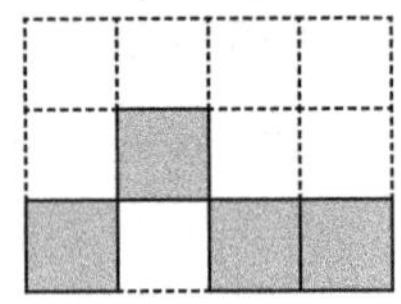

Inquisitive ant

circumference
The distance around the edge of a circle.

Issue 26

Measurement

Prerequisites for learning

- Read, write and convert time between analogue and digital 12- and 24-hour clocks
- Begin to understand compound units for speed, such as kilometres per hour
- Calculate mentally with integers and decimals
- Use formal written methods to add, subtract, multiply and divide integers and decimals
- Understand the concept of volume

Resources

pencil and paper
Resource sheet 2: My notes (optional)
Resource sheet 3: Pupil self assessment booklet (optional)
Resource sheet 18: 1 cm squared paper
ruler
calculator
analogue clock with geared hands (optional)
stopwatch
data handling software (optional)
boxes
newspaper
sticky tape
collection of cylinders
different types of small objects such as counters, paper clips, Centicubes, marbles
computer with Internet access

Teaching support

Page 1

Sports Update

- There will be other activities on the children's lists that are not in the table. Ask children to estimate, or find out on the Internet, how much energy each of these uses per minute.

Let's Investigate

- 12-hour digital clock:
  57 palindromic times in 12 hours
  114 palindromic times in 24 hours (57 × 2)

12.21	1:01	2:02	3:03	4:04	5:05	6:06	7:07	8:08	9:09	10:01	11:11
	1:11	2:12	3:13	4:14	5:15	6:16	7:17	8:18	9:19		
	1:21	2:22	3:23	4:24	5:25	6:26	7:27	8:28	9:29		
	1:31	2:32	3:33	4:34	5:35	6:36	7:37	8:38	9:39		
	1:41	2:42	3:43	4:44	5:45	6:46	7:47	8:48	9:49		
	1:51	2:52	3:53	4:54	5:55	6:56	7:57	8:58	9:59		

- 24-hour digital clock:
  16 palindromic times

  01:10 02:20 03:30 04:40 05:50 10:01 11:11 12:21 13:31 14:41 15:51 20:02
  21:12 22:22 23:32 24:42

Page 2

Around the World

- 1000 hours is 41 days and 16 hours.
- 41 days later than 19:30 on Thursday 14th July is 19:30 on Wednesday 24th August.
- Another 16 hours is 11:30 on Thursday 25th August – six weeks later.

Around the World

- Ensure the children realise that their comparative statements will be based on approximations. However, children should be able to show justifications for their approximations.

Around the World

- No aeroplane flies directly from London to Sydney, they all have to stop somewhere en route, and the distance can vary significantly depending on where this is. Also, the flight paths they use are not always the most direct route.
- The flight time is around 21·5 hours.
- A Boeing 747 flies at approximately 900 km/h, or 15 km/min.

Page 3

Focus on Science

- Children present the results of their experiment in a table or graph using ICT if appropriate.

Construct

- Finding the surface area of the cuboid will help children find out how much wrapping paper is needed. This is calculated by adding together the area of each face of the cuboid to find the total surface area.
- Ask children to find a formula for the size of paper needed to wrap any cuboid object, i.e. $2(hw + hd + hd + d^2)$, where h = height of cuboid, w = width of cuboid, d = depth of cuboid and d^2 = overlap.

Let's Investigate

- If children are quick to appreciate that the volume of a cylinder is found by multiplying the area of the circle (the base) by the length of the cylinder, then discuss with them the formula for working out the volume of a cylinder, i.e. Area of a circle = πr^2. So, volume of a cylinder of length $l = \pi r^2 \times l$ (or $\pi r^2 l$).

Page 4

Construct

- Rather than having each child design four floor plans (one for each sized flat), you may wish to ask individual children to just draw a plan for one of the types of accommodation. If so, ask different children to design plans for each of the type of accommodation.

Money Matters

- Children will need to decide what price per square metre to charge. This will vary enormously, so an important part of their answers is the reasoning they offer for their pricing.

AfL

- What did you find out about this investigation?
- How did you start this investigation?
- What calculations did you do?
- Talk me through your experiment. If you were devising a similar experiment again, what would you do differently?
- What things did you think about as you were planning your designs? Why were these important considerations?
- What was your reasoning for making these decisions? Why are they valid reasons?

Answers

Page 1

Sports Update
Answers will vary.

Let's Investigate
12-hour digital clock: 57 palindromic times in 12 hours
(114 palindromic times in 24 hours)
24-hour digital clock: 16 palindromic times

Page 2

Around the World
11:30 on Thursday 25th August.

Around the World
Commercial airliners fly at around 10 000 m altitude.
Smaller planes without pressurised cabins can fly at around 3000 m – 4000 m altitude.
The Eiffel Tower is 324 m high; the London Eye is 135 m high.

Around the World
It is 16 980 km from London to Sydney.
21·5 hours flight time.
Flight speed is around 900 km/h.
Answers will vary.

Page 3

Focus on Science
Results will vary.

Construct
Answers will vary.

Let's Investigate
To find the volume of the cylinder, multiply the area of the circle (the base) by the length of the cylinder.

Page 4

Construct
Designs will vary.

Money Matters
Answers will vary.

Inquisitive ant

metrology
The scientific study of units of measurement.

Issue 27

2-D shapes

Prerequisites for learning

- Recognise and use square numbers and the square roots of perfect squares
- Compare and classify geometric shapes, including quadrilaterals and triangles, based on their properties and sizes; use knowledge of properties to draw 2-D shapes
- Draw and measure lines to the nearest millimetre

Resources

pencil and paper
Resource sheet 2: My notes (optional)
Resource sheet 3: Pupil self assessment booklet (optional)
Resource sheet 7: Pythagoras's Theorem
Resource sheet 8: Seven circles patterns
Resource sheet 18–20: 1 cm / 2 cm / 0·5 cm squared paper
Resource sheet 21: Squared dot paper (optional)
ruler
material for making a poster
geoboards and elastic bands
matchsticks
computer

Teaching support

Page 1

Famous Mathematicians

- Ensure children are familiar with the term *hypotenuse* (c), meaning the longest side of a right-angled triangle, opposite the right angle.
- In this and the next activity, children are introduced to Pythagoras and his theorem. Both these practical activities should help children realise that for any right-angled triangle, the square of the length of the hypotenuse is equal to the sum of the squares of the lengths of the other two sides, i.e. $c^2 = a^2 + b^2$.

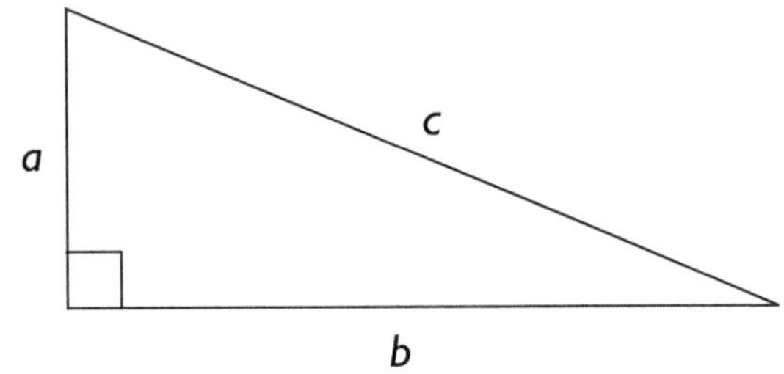

The Puzzler

- Children can either use blank paper or 1 cm, 2 cm or 0·5 cm squared paper to draw the square.

Famous Mathematicians

- See Famous Mathematicians comment above.

Page 2

The Language of Maths

- Provide the children with a list of quadrilaterals for them to investigate.
- Ensure the children include an *oblong* in their list of quadrilaterals i.e. a rectangle that is not a square.

What's the Problem?

- This is a very open-ended problem where different children are more than likely to have completed different correct grids. When children have completed the activity, provide an opportunity for them to talk about their grid to the rest of group. Discuss the similarities and differences between grids.

Let's Investigate

- Encourage the children to draw not only the shapes they make using the two squares, but how they are made from the two squares, i.e.

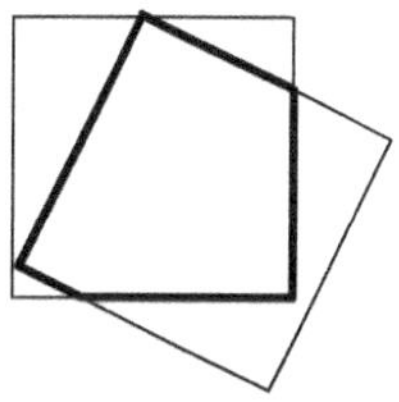

rather than

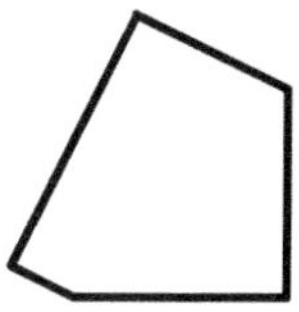

Page 3

Let's Investigate

- Provide the children with geoboards and elastic bands to make the shapes before drawing them onto squared paper.
- Tell the children that it is possible to make 16 quadrilaterals on a 3 × 3 dot grid (see Answers) including:
  - two different squares
  - one rhombus
  - one rectangle / oblong
  - two different parallelograms
  - three trapeziums
  - one kite
  - two different inverted kites (delta)
  - four different irregular quadrilaterals

Let's Investigate

- Explain to the children that they need to be as accurate as possible in their measuring and drawing of each of the shapes, and in measuring the midpoints of each side of the quadrilaterals.
- If appropriate, provide the children with 1 cm, 2 cm or 0·5 cm squared paper.

Let's Investigate

- Ensure children realise that to make a 3 × 3 square on squared dot paper they need to use a total of 16 dots (4 × 4).
- Do children recognise the relationship between the total number of dots in the square and the polygon that can be made with the greatest number of sides? Can they use and apply this relationship when drawing polygons on a 4 × 4 or 5 × 5 square?

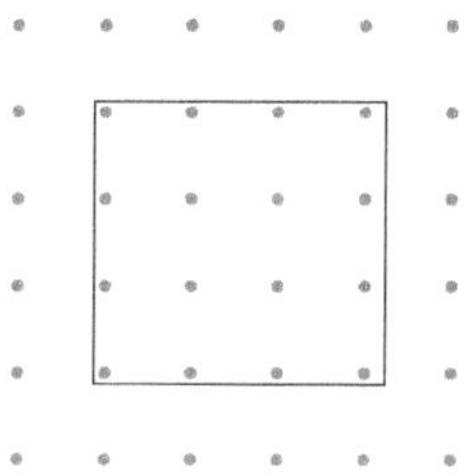

Page 4

Construct

- Children can either use blank paper or 1 cm, 2 cm or 0·5 cm squared paper to draw their shapes.
- Encourage the children to show in their drawings which shapes have been joined together to make each of the different polygons, i.e.

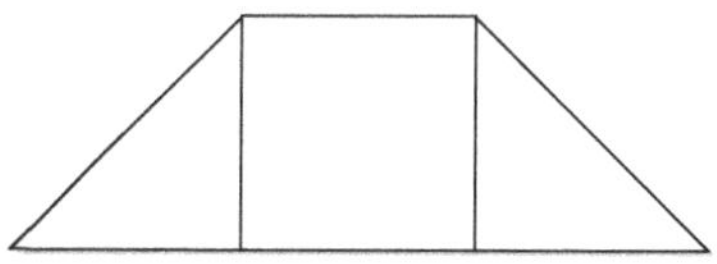

Let's Investigate

- Suggest the children draw the polygons on squared dot paper to better show the matchsticks.

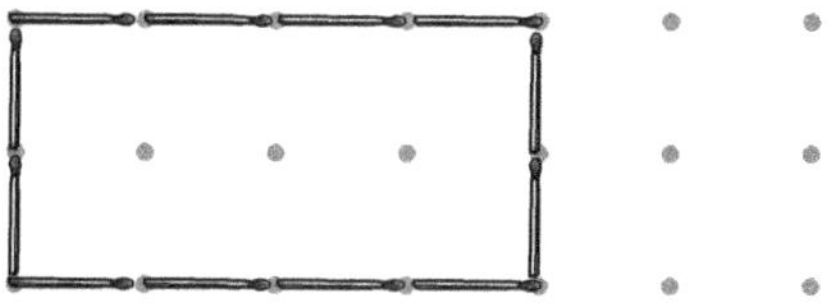

Let's Investigate

- For this activity, most children will probably require more than one copy of Resource sheet 8: Seven circles patterns.

AfL

- Explain the formula to me in your own words.
- Tell me the name of a special quadrilateral. What are its properties?
- What did you find out as a result of your investigation? What patterns / relationships did you notice?
- What shapes were you able to make? How?

Answers

Page 1

Famous Mathematicians
The area of the square on side a of the triangle is a^2. The area of the square on side b of the triangle is b^2. The area of the square on side c of the triangle is c^2.
If you add up the total number of squares in square a and square b, it is the same number as in square c.

The Puzzler
Using three pieces, the following shapes can be made: right-angled triangle, square, rectangle, parallelogram, trapezium, pentagon.

Using two pieces, the following shapes can be made: right-angled triangle, trapezium, pentagon.

Famous Mathematicians

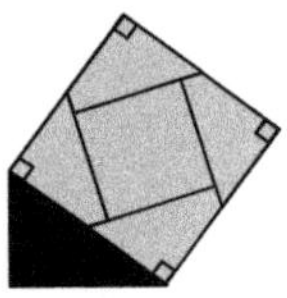

Page 2

The Language of Maths
A quadrilateral is a four-sided shape. There are six special quadrilaterals:

parallelogram: 2 pairs of opposite parallel sides / opposite sides and angles are equal

rhombus: a special parallelogram / 4 equal sides / opposite sides equal and parallel / opposite angles equal / 2 lines of symmetry

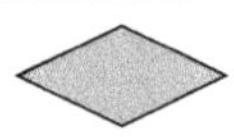

rectangle: a special parallelogram / 4 right angles / opposite sides equal and parallel / opposite angles equal / 2 lines of symmetry

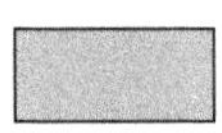

square: a special parallelogram and a special rhombus / 4 equal sides / 4 right angles / opposite sides equal and parallel / 4 lines of symmetry

kite: 2 pairs of adjacent sides equal / 1 pair of opposite angles equal / 1 line of symmetry

trapezium: 1 pair of parallel sides

What's the Problem?
Drawings and naming will vary. However, most grids should show shapes similar to the following:

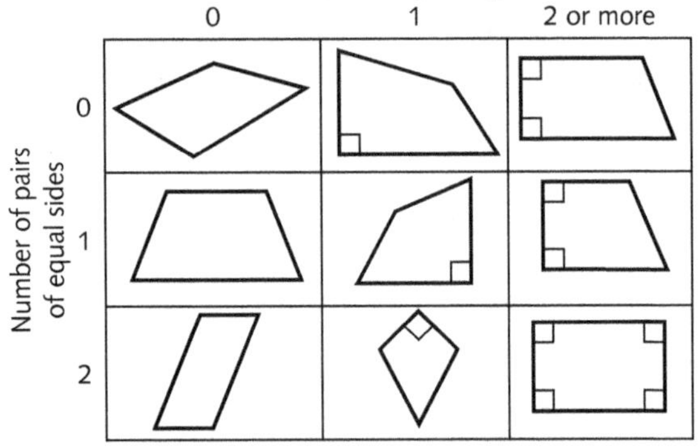

Page 3 (continued)

Let's Investigate
It is possible to make a rectangle, kite, isosceles triangle, hexagon and octagon in the same way.

It is not possible to make a shape with more than eight sides.

With three squares it is also possible to make a trapezium and shapes with up to twelve sides.

Page 3

Let's Investigate

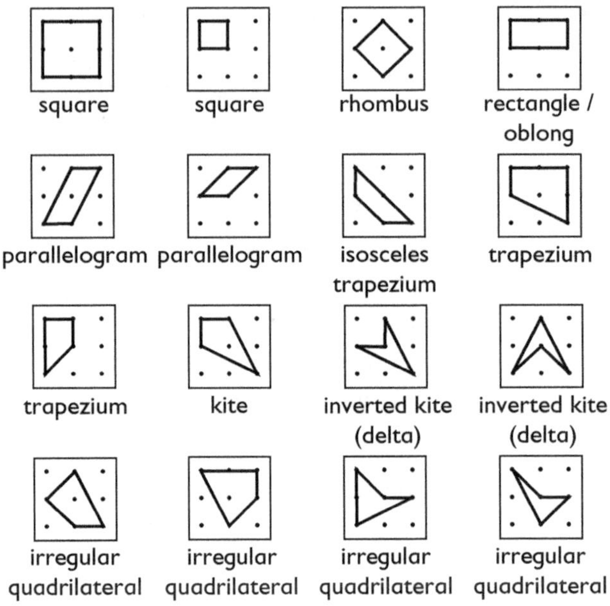

Let's Investigate
When the midpoints of adjacent sides of any quadrilateral are joined together, a parallelogram is formed.

Let's Investigate
Answers will vary.

It is possible to draw a 16-sided polygon on a 3 × 3 square. For example:

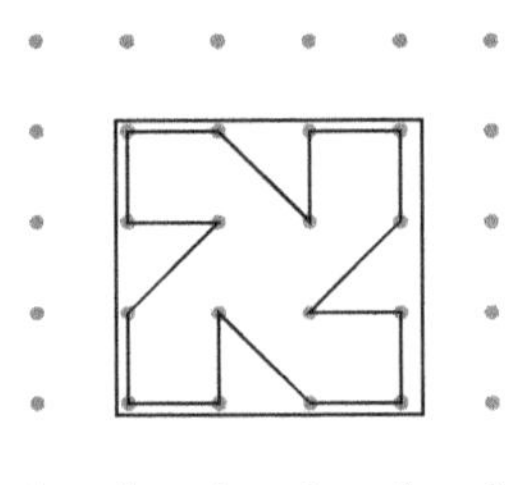

It is possible to draw a 25-sided polygon on a 4 × 4 square.

It is possible to draw a 36-sided polygon on a 5 × 5 square.

Page 4

Construct
Answers will vary.

Let's Investigate
Answers will vary.
Each polygon will have a perimeter of 12 matchsticks.

Let's Investigate
Results of the investigation will vary.

Inquisitive ant

icosagon
A polygon with 20 sides.

Issue 28

3-D shapes

Prerequisites for learning

- Identify, visualise and describe properties of 3-D shapes
- Identify and draw nets of 3-D shapes
- Understand the concepts of volume and surface area of cubes

Resources

pencil and paper
Resource sheet 2: My notes (optional)
Resource sheet 3: Pupil self assessment booklet (optional)
Resource sheet 18: 1 cm squared paper
Resource sheet 19: 2 cm squared paper
Resource sheet 23: Isometric paper
set of geometric 3-D shapes with a wide selection of different polyhedrons (optional)
apparatus used for constructing 3-D shapes (optional)
ruler
material for making a poster
coloured pencils
Centicubes
computer with Internet access

Teaching support

Page 1

Let's Investigate

- A truncated cube is one of the 13 Archimedean solids.
- Ask the children to investigate other Archimedean solids.

Let's Investigate

- 5 ones, 1 two – 1 way
  4 ones, 2 twos – 2 ways
  3 ones, 3 twos – 2 ways
  2 ones, 4 twos – 2 ways
  1 one, 5 twos – 1 way
- 4 ones, 1 two, 1 three – 2 ways
  3 ones, 1 two, 2 threes – 3 ways
  3 ones, 2 twos, 1 three – 3 ways
  2 ones, 1 two, 3 threes – 3 ways
  2 ones, 2 twos, 2 threes – 5 ways
  2 ones, 3 twos, 1 three – 3 ways
  1 one, 4 twos, 1 three – 2 ways
  1 one, 3 twos, 2 threes – 3 ways
  1 one, 2 twos, 3 threes – 3 ways
  1 one, 1 two, 4 threes – 2 ways

The Puzzler

- The important aspect of this activity is the justifications children give for their reasoning.
- Have available a set of geometric 3-D shapes that include a wide selection of different polyhedrons.

Page 2

Construct

- Children should not have difficulty in designing and constructing a cuboid.
- If available, provide the children with apparatus used for constructing 3-D shapes.

Construct

- If available, provide the children with apparatus used for constructing 3-D shapes.
- Children plan, draw and construct nets of other 3-D shapes.

The Language of Maths

- Children will need to have access to the Internet and a printer for this activity. Suggest the children create a poster of their images.
- Children compare Archimedean solids and Platonic solids.

Page 3

The Language of Maths

- Children will need to have access to the Internet and a printer for this activity. Suggest the children create a poster of their images.

Construct

- Tell children to draw all the front faces first, then the side faces and finally the top faces.
- Suggest that children make the two shapes out of Centicubes before drawing them.

Page 4

Looking for Patterns

- To assist children in identifying the pattern, tell them to find the differences between adjacent numbers going down each column.

Construct

- If appropriate, suggest the children design their nets using ICT.
- Drawing tabs on the net will ensure that the 3-D shape holds together better. Ask the children to work out the fewest number of tabs they would need to include in their nets and where these should be located.

AfL

- What can you tell me about truncated solids?
- What 3-D shapes are these? How do you know?
- How did you know what shape the net needed to be?
- What is a perfect polyhedron? What are the special names given to each of them?
- What is a prism?
- Was your net successful? What would you do next time to improve your net?
- What did you discover as a result of your investigation? What patterns did you notice? What generalisations can you make?

Answers

Page 1

Let's Investigate
A truncated cube has 14 faces (8 triangles, 6 octagons), 36 vertices and 24 edges.

Let's Investigate
8 arrangements of ones and twos are possible.
29 arrangements of 1, 2 and 3 are possible.

The Puzzler
Sphere, cone, triangular-based pyramid / hexahedron, square-based pyramid / octahedron, cuboid, hexagonal prism, pentahedron / decahedron / nonahedron / tetradecahedron.

Page 2

Construct

Nets will vary.

Construct

Nets will vary.

The Language of Maths
Tetrahedron, cube, octahedron, dodecahedron and icosahedron.
Images will vary.

Page 3

The Language of Maths
A prism is a 3-D shape with a constant cross-section; both ends of the solid are the same and parallel, and anywhere it is cut parallel to the ends gives the same shape.
Images will vary.

Construct

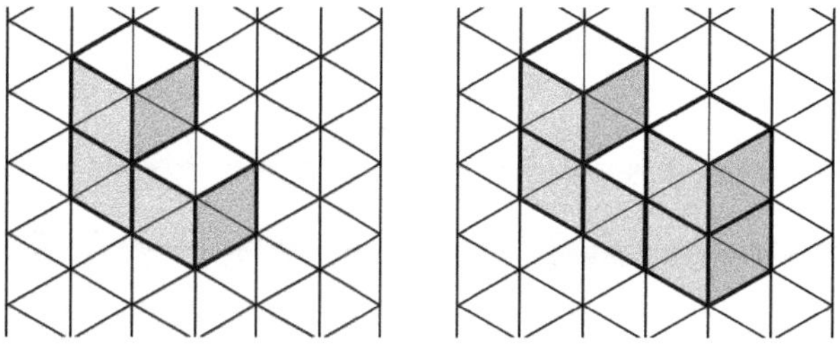

Page 4

Looking for Patterns

Size of cube	Face area	Surface area	Volume
1 cm	1 cm^2	6 cm^2	1 cm^3
2 cm	4 cm^2	24 cm^2	8 cm^3
3 cm	9 cm^2	54 cm^2	27 cm^3
4 cm	16 cm^2	96 cm^2	64 cm^3
5 cm	25 cm^2	150 cm^2	125 cm^3
6 cm	36 cm^2	216 cm^2	216 cm^3
7 cm	49 cm^2	294 cm^2	343 cm^3
8 cm	64 cm^2	384 cm^2	512 cm^3
9 cm	81 cm^2	486 cm^2	729 cm^3
10 cm	100 cm^2	600 cm^2	1000 cm^3

Patterns will vary. They may include:
Face area:
- Square numbers
- The difference between consecutive face areas starts at 3 cm^2 and increases by 2 cm^2 each time, i.e. 3 cm^2, 5 cm^2, 7 cm^2, 9 cm^2, 11 cm^2, 13 cm^2, 15 cm^2, 17 cm^2, 19 cm^2.

Surface area:
- The difference between consecutive surface areas starts at 18 cm^2 and increases by 12 cm^2 each time, i.e. 18 cm^2, 30 cm^2, 42 cm^2, 54 cm^2, 66 cm^2, 78 cm^2, 90 cm^2, 102 cm^2, 114 cm^2.

Volume:
- Cubed numbers

Construct
Nets will vary.

Inquisitive ant

trihedral
Having three plane faces.

Issue 29

Movement and angle

Prerequisites for learning

- Know angles are measured in degrees: estimate and compare acute, obtuse and reflex angles
- Draw given angles, and measure them in degrees (°)
- Identify:
  - angles at a point and one whole turn (total 360°)
  - angles at a point on a straight line and $\frac{1}{2}$ a turn (total 180°)
  - other multiples of 90°
- Find unknown angles in any triangles, quadrilaterals, and regular polygons
- Understand and use simple algebraic conventions
- Use and construct simple expressions and formulae in words and symbols

Resources

pencil and paper
Resource sheet 2: My notes (optional)
Resource sheet 3: Pupil self assessment booklet (optional)
Resource sheet 9: Analogue clocks
Resource sheet 10: Polygons
Resource sheet 21: Squared dot paper
Resource sheet 24: Diamond paper
ruler
protractor
A4 paper
scissors
computer

Teaching support

Page 1

Let's Investigate

- This activity, and the first Let's Investigate activity on page 2, are both concerned with finding different ways of travelling along grid lines between two points. Suggest the children write instructions for moving between three or more points, including instructions for moving back to the starting point, e. g. Point A, Point B, Point C and Point A.

Construct

- Ensure that children are familiar with the convention of writing the size of an angle inside an arc, and drawing a small square to indicate a right angle, i.e.

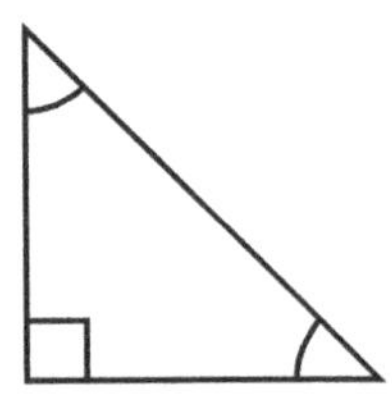

- Remind children that all the angles in a triangle add up to 180°.
- Suggest children draw triangles with multiples of 5°, e.g. 45°, 60° and 75°.

Let's Investigate

- Remind children that angles on a straight line total 180°.

Page 2

Let's Investigate

- This activity, and the first Let's Investigate activity on page 1, are both concerned with finding different ways of travelling along grid lines between two points. Suggest the children write instructions for moving between three or more points, including instructions for moving back to the starting point.

Let's Investigate

- Ensure children realise that the angle formed by the two hands of a clock can be measured in two ways – an acute / obtuse angle or a reflex angle. Encourage children to measure the acute / obtuse angle.
- When children have made up different times and measured the angles, ask them to arrange their times into three groups. Times where the hands form:
  - an acute angle
  - a right angle
  - an obtuse angle.

Page 3

Let's Investigate

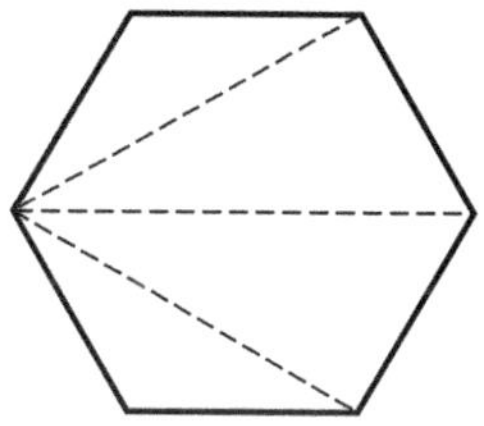

- If you draw all the diagonals possible from one vertex in a polygon with four or more sides, the polygon is then divided into several non-overlapping triangles, e.g.
  The interior angle sum of this polygon can now be found by multiplying the number of triangles by 180°.
- By investigating different polygons, it can be found that the number of triangles is always two less than the number of sides. This fact is stated in the formula.
- If n represents the number of sides of a polygon and the number of triangles that can be made is $n - 2$, then the formula for calculating the sum (S) of interior angles is $S = 180 \times (n - 2)$.

Let's Investigate

- Children need to realise that:
  - there are 360° in the circle of an analogue clock and that this is divided into 12 equal sectors, each of 30°
  - both hands on an analogue clock move at a constant rate – in an hour, the minute hand moves through 360° and the hour hand through 30°
  - at 15 minutes past the hour, the hour hand is $\frac{1}{4} \times 30°$ past the hour division, i.e. 7·5°
  - at 30 minutes past the hour, the hour hand is $\frac{1}{2} \times 30°$ past the hour division, i.e. 15°
  - at 45 minutes past the hour, the hour hand is $\frac{3}{4} \times 30°$ after the hour division, i.e. 22·5°.
- Provide children with a copy of Resource sheet 9: Analogue clocks, and allow them to draw the hands on the clocks.

Page 4

Let's Investigate

- Ask the children to sort the letters into different groups, e.g.
  - right angles only
  - acute angles only
  - obtuse angles only
  - right angles and acute angles
  - right angles and obtuse angles
  - acute angles and obtuse angles
  - right angles, acute angles and obtuse angles
- Children also investigate reflex angles in different capital letters.

Let's Investigate

- Encourage the children to look beyond the more obvious examples when finding different types of angles.

Let's Investigate

- Children need to understand the algebraic convention for multiplication, e.g. $6m = 6 \times m$ and $\frac{1}{2}m = \frac{1}{2} \times m$.
- Provide children with a copy of Resource sheet 9: Analogue clocks, and allow them to draw the hands on the clocks.
- Ask the children to use the formula to calculate times, other than 12 o'clock, when the hands of the clock are on top of each other (i.e. when $5\frac{1}{2}m - 30h = 0$).

AfL

- Describe the route to me.
- What instructions did you write?
- Did you find another route?
- How did you work out / calculate what the size of the angle was?
- What have you found out about the different angles formed by the hands on a clock / letters of the alphabet?
- Tell me some objects that show examples of right angles. What about acute angles? Obtuse angles? Reflex angles?

Answers

Page 1

Let's Investigate
Right 1, down 1, right 1, down 1
Down 1, right 2, down 1
All these routes are the same length and the shortest way from A to E.

Construct
60°, 80° and 40°
90°, 40° and 50°
30°, 80° and 70°
40°, 40° and 100°

Let's Investigate
Answers will vary.

Opposite angles of a parallelogram are equal; adjacent angles total 180°. Sum of the internal angles is 360°.

Sum of the internal angles of a trapezium is 360°. An isosceles trapezium has two pairs of equal angles; the sum of one of each pair of angles is 180°.

Page 2

Let's Investigate
Down right 2, down left 2
Down right 1, down left 2, down right 1
Down left 1, down right 1, down left 1, down right 1
Down right 1, down left 1, down right 1, down left 1
Answers will vary.

Let's Investigate
The angle formed by the hands at 11:35 is 137° or 138°.
Answers will vary.

Page 3

Let's Investigate
If n represents the number of sides of a polygon and the number of triangles that can be made is $n - 2$, then the formula for calculating the sum (S) of interior angles is $S = 180 \times (n - 2)$.

Let's Investigate

	o'clock	15 min past	30 min past	45 min past
12	0°	82·5°	165°	112·5°
3	90°	7·5°	75°	157·5°
6	180°	97·5°	15°	67·5°
9	90°	172·5°	105°	22·5°

Page 4

Let's Investigate
Answers will vary, but should include the letters A, E, F, H, K, L, M, N, R, T, V, W, X, Y, Z.

Let's Investigate
Results of the investigation will vary.

Let's Investigate
When using the formula to check the answers to the second Let's Investigate activity on page 3 children will twice (12:45 and 9:15) obtain the reflex angle rather than the acute/obtuse angle.
Answers will vary.

Inquisitive ant

theodolite
An instrument consisting of a rotating telescopic sight, used by surveyors to measure horizontal and vertical angles.

Issue 30

Reflections

Prerequisites for learning

- Recognise reflective symmetry
- Complete patterns with up to two lines of symmetry
- Identify, describe and represent the position of a shape following a reflection, using the appropriate language, and know that the shape has not changed
- Use all four quadrants to find coordinates of points determined by geometric information

Resources

pencil and paper
Resource sheet 2: My notes (optional)
Resource sheet 3: Pupil self assessment booklet (optional)
Resource sheet 8: Seven circles patterns
Resource sheet 11: Coordinates grids (0 to ±4)
Resource sheet 18: 1 cm squared paper
Resource sheet 19: 2 cm squared paper
examples of cross-stitching (optional)
ruler
mirror (optional)
material for making a poster
coloured pencils
calculator
ink and / or paints
poster paper
computer with Internet access

Teaching support

Page 1

Construct

- If possible, show children an example of cross-stitching.
- If appropriate, children can create their cross-stitch needlework pictures or designs using ICT.

Construct

- You may need to vary this activity according to the specific circumstances of your school.

Let's Investigate

- Ensure children understand the difference between a logo and a company or product name.

Page 2

Let's Investigate

- If children have already completed Issue 32 – Reflective and rotational symmetry, ask them to comment on the similarities and differences between the patterns they create here and those in the Let's Investigate activity on page 2 of Issue 32 – Reflective and rotational symmetry.

The Puzzler

- Children should be encouraged to complete this activity without the use of a mirror.
- Note that the 1 digit is only symmetrical when it is with other 1 digits. This is because it is displayed on the right-hand side of the seven-segment display, not in the middle.
- Provide the children with a mirror.
- Explain to the children how the only digits that display reflective symmetry on a calculator are 0, 2, 5 and 8.
- Some children may need to be shown how the mirror image of the digit 2 on a calculator appears as the digit 5, and vice versa.

Let's Investigate

- Ensure that children are able to read and plot points in all four quadrants to find coordinates of points determined by geometric information.

Page 3

Let's Investigate

- Draw children's attention to the link between the second example here and the second Let's Investigate activity on page 2.

Focus on Science

- Encourage children to look for examples of reflective symmetry in plants and animals as well.

Page 4

Construct

- Be aware that this activity is likely to be rather 'messy'. However, it is worth allowing time for the children to create their ink blots. It should not take them too long to do this, but it is recommended that the children make their ink blots during one session and write their descriptions during another session.

Let's Investigate

- Children will need to have access to the Internet and a printer for this activity. Ensure children annotate their images of rugs and carpets, commenting on the symmetry in each example.

AfL

- Describe to me the symmetry in your pattern / logo.
- How does this show reflective symmetry?
- What type of symmetry is this?
- What generalisations can you make about reflective symmetry?
- What strategies did you use to find the answer to this puzzle? How did they help?
- Describe to me the symmetry in these natural objects / ink blots / rugs.

Answers

Page 1

Construct
Pictures and designs will vary.

Construct
Logos will vary.

Let's Investigate
Results of the investigation will vary.

Page 2

Let's Investigate
Patterns will vary.

The Puzzler
Seven three-digit numbers:

111	205	502	808
	285	582	888

Thirteen four-digit numbers:

1111	2005	5002	8008
	2255	5252	8258
	2525	5522	8528
	2885	5882	8888

Let's Investigate

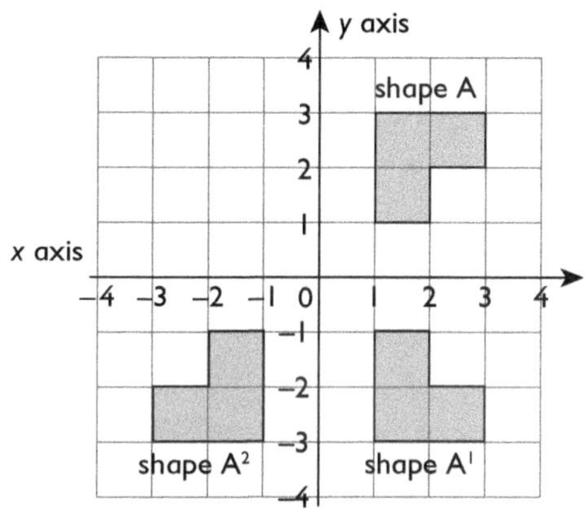

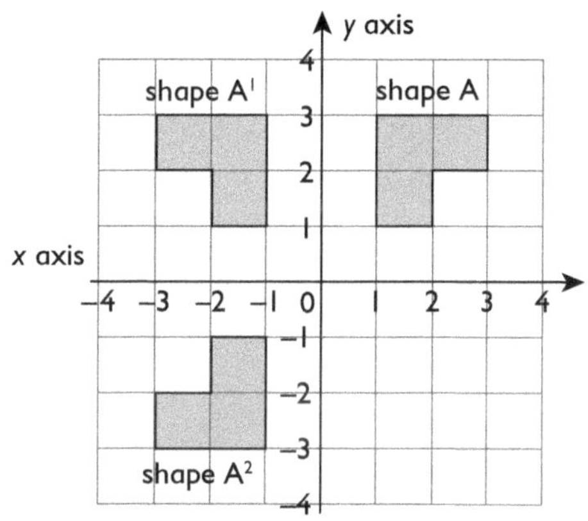

Both Shapes A^{2}s are the same.

Page 3

Let's Investigate
Both statements are true.

Focus on Science
Results of the investigation will vary.

Page 4

Construct
Ink blots and descriptions will vary.

Let's Investigate
Results of the investigation will vary.

Inquisitive ant

axis of symmetry (also referred to as line of *symmetry* or *mirror line*)
The line around which a geometrical shape is symmetrical.

Issue 31

Translations

Prerequisites for learning

- Identify, describe and represent the position of a shape following a translation, using the appropriate language, and know that the shape has not changed
- Understand tessellations
- Describe and identify the position of a square on a grid of squares
- Use coordinates in all four quadrants to draw, locate and complete shapes that meet given properties

Resources

pencil and paper
Resource sheet 2: My notes (optional)
Resource sheet 3: Pupil self assessment booklet (optional)
Resource sheet 11: Coordinates grids (0 to ±4)
Resource sheet 12: Translating squares
Resource sheet 18: 1 cm squared paper
Resource sheet 19: 2 cm squared paper
Resource sheet 24: Diamond paper
ruler
two coins, each a different denomination
1–6 dice
scissors
poster paper
range of colouring materials
examples of cross-stitching (optional)
computer with Internet access

Teaching support

Page 1

Construct

- Ensure the children realise that they are always translating the original shape they drew in the centre of the sheet.
- Discuss with the children a rule for what happens when they reach the edge of the paper – do they roll the dice and spin the coin again?

Looking for Patterns

- The important aspect of this activity is the translating pattern children make after they have arranged the four small squares into one larger square pattern.
- Children record their translating patterns on 2 cm squared paper.

Page 2

Construct

- If possible, show children an example of cross-stitching.
- If appropriate, children can create their cross-stitch needlework pictures or designs using ICT.

The Language of Maths

- Ensure that children are able to read and plot points in all four quadrants to find coordinates of points determined by geometric information.

In the Past

- This activity looks at the work of William Morris and how different types of transformations featured in his work. Ensure children realise that they need to comment on reflections, rotation and translations.
- Children will need to have access to the Internet and a printer for this activity.
- Suggest that the children make a display of the work of William Morris. Ensure that they annotate their images, commenting on the transformations in each example.

Page 3

Looking for Patterns

- The reason for asking the children to suggest different ways of translating each shape is to reinforce the notion that with translations it does not matter in which order two translations are done, the end result will always be the same. For example, shape A^1 has been translated 4 squares up and 2 squares to the left, or 2 squares to the left and 4 squares up.

The Arts Roundup

- There are two aspects to this investigation. Firstly, investigating the work of Piet Mondrian, and secondly designing a Piet Mondrian inspired work that includes examples of translations.

Around the World

- There are two aspects to this investigation. Firstly, investigating Aboriginal art, and secondly designing an Aboriginal artwork that includes examples of translations.

Page 4

Around the World

- This activity requires the children to describe and identify the position of a set of squares on a grid of squares and then translate these sets.
- Children write their own set of instructions for translating each of the ships on the grid. They then swap their instructions and translate each of the ships accordingly.

Around the World

- Children write an alternative route for the sailing boat.
- What is the quickest route the sailing boat can take?

AfL

- Explain your pattern to me. How does it show translations?
- How has this shape been translated?
- What happened when you translated your shape?
- Can you tell me some things that show a translating pattern? How do they show it?
- What did you find out about the work of William Morris / Piet Mondrian?
- How does your picture show examples of translations?
- What are the new positions of the ships? How have they been moved?
- Describe for me the route of the sailing boat. Is there another route the boat could have taken? Which is the quickest route? Why?

Answers

Page 1

Construct
Patterns will vary.

Looking for Patterns
Patterns will vary.

Page 2

Construct
Patterns will vary.

The Language of Maths
A^1 translated to:
(3,–3), (3,–2), (4,–2), (4,–1), (1,–1), (1,–2), (2,–2), (2,–3), (3,–3)

A^2 translated to:
(–2,–3), (–2,–2), (–1,–2), (–1,–1), (–4,–1), (–4,–2), (–3,–2), (–3,–3), (–2,–3)

In the Past
Results of the investigation will vary.

Page 3

Looking for Patterns
A^2 4 squares up and 5 squares to the right or 5 squares to the right and 4 squares up.
A^3 3 squares up and 2 squares to the right or 2 squares to the right and 3 squares up.
A^4 1 square up and 5 squares to the left or 5 squares to the left and 1 square up.
A^5 1 square down and 3 squares to the right or 3 squares to the right and 1 square down.
A^6 2 squares down and 3 squares to the left or 3 squares to the left and 2 squares down.
A^7 3 squares down and 5 squares to the right or 5 squares to the right and 3 squares down.
A^8 4 squares down and 1 square to the left or 1 square to the left and 4 squares down.
A^9 5 squares down and 5 squares to the left or 5 squares to the left and 5 squares down.
A^{10} 5 squares down and 2 squares to the right or 2 squares to the right and 5 squares down.

The Arts Roundup
Results of the investigation will vary.

Around the World
Results of the investigation will vary.

Page 4

Around the World

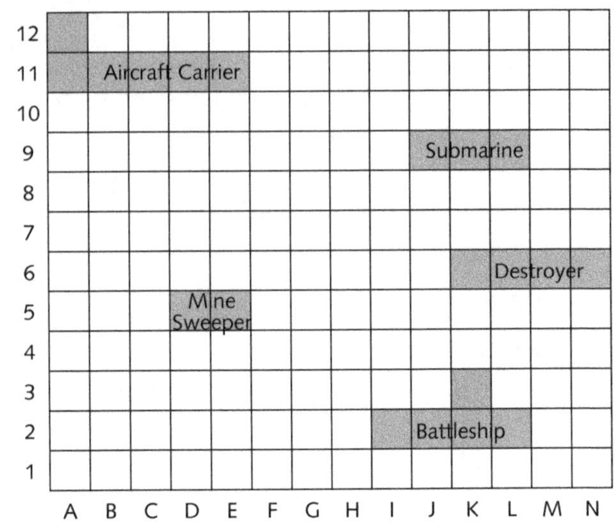

Around the World
Answers will vary.

Inquisitive ant

tessellation
A pattern made by repeated use of the same shape fitting together without overlapping or leaving any spaces.

Issue 32

Reflective and rotational symmetry

Prerequisites for learning

- Recognise and use quarter turns, both clockwise and anticlockwise and know that a right angle represents a quarter turn
- Know angles are measured in degrees (°)
- Identify:
  - angles at a point and one whole turn (total 360°)
  - angles at a point on a straight line and $\frac{1}{2}$ a turn (total 180°)
  - other multiples of 90°
- Draw the position of a shape after a rotation

Resources

pencil and paper
Resource sheet 2: My notes (optional)
Resource sheet 3: Pupil self assessment booklet (optional)
Resource sheet 8: Seven circles patterns
Resource sheet 18: 1 cm squared paper
Resource sheet 19: 2 cm squared paper
Resource sheet 21: Squared dot paper
Resource sheet 22: Triangular dot paper
examples of cross-stitching (optional)
ruler
kaleidoscope (optional)
coloured pencils
material for making a poster
cardboard
scissors
A4 paper
camera
different packs of playing cards (optional)
computer with Internet access

Teaching support

Page 1

Construct

- If possible, show children an example of cross-stitching.
- If appropriate, children can create their cross-stitch needlework pictures or designs using ICT.

Technology Today

- If appropriate, children should be encouraged to show the rotation of each shape, and the position of the star, using ICT.
- Alternatively, provide children with squared and triangular dot papers to assist them with drawing the shapes.

The Language of Maths

- Most children should be familiar with kaleidoscopes, but it would be helpful to have one for the children to use, and if possible, opened up to show the workings.

Page 2

Let's Investigate

- If children have already completed Issue 30 – Reflections, ask them to comment on the similarities and differences between the patterns they create here and those in the first Let's Investigate activity on page 2 of Issue 30 – Reflections.

Around the World

- Children can write their description using ICT.

Around the World

- Children investigate other buildings that show examples of different transformations.

Page 3

Looking for Patterns

- Children can generate their tile using ICT. They then use the tile to make a rotational design.

Looking for Patterns

- Children investigate rotating different regular polygons 180° about the midpoint of one of its sides, e.g.

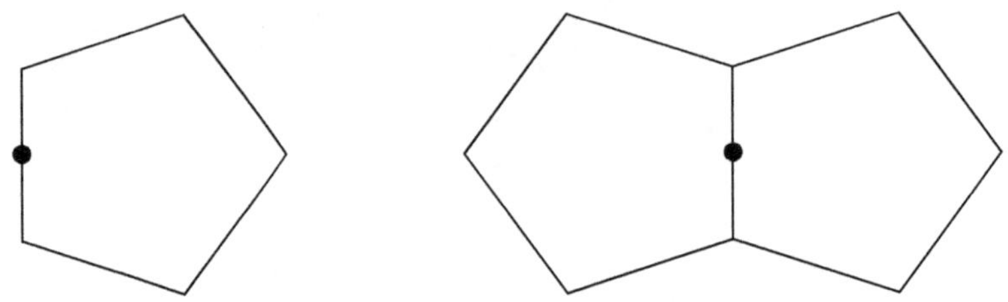

Focus on Science

- Encourage children to fold their paper into quarters, eighths etc. before cutting.
- Ask children to fold the paper in different ways, e.g. diagonally, and comment on the number of rotational symmetries. What if they use different shapes of paper, e.g. circular?

Page 4

Let's Investigate

- Rather than drawing the plug holes, allow the children to use the school camera to take photos of them.

The Language of Maths

- How does the font used to write a letter affect whether or not the letter has rotational symmetry?

The Language of Maths

- Try and provide the children with as many different sets of playing cards as possible.

AfL

- Explain your pattern to me. How does it show rotational symmetry?
- How have the solids been rotated?
- Can you tell me some things that show rotational symmetry? How do they show it?
- What did you find out about the patterns in The Alhambra / snowflakes / plug holes? How are they symmetrical?
- Which letters of the alphabet show rotational symmetry?
- Why do you think playing cards have rotational symmetry? How do your digit cards show rotational symmetry?

Answers

Page 1

Construct
Pictures and designs will vary.

Technology Today
Answers will vary.

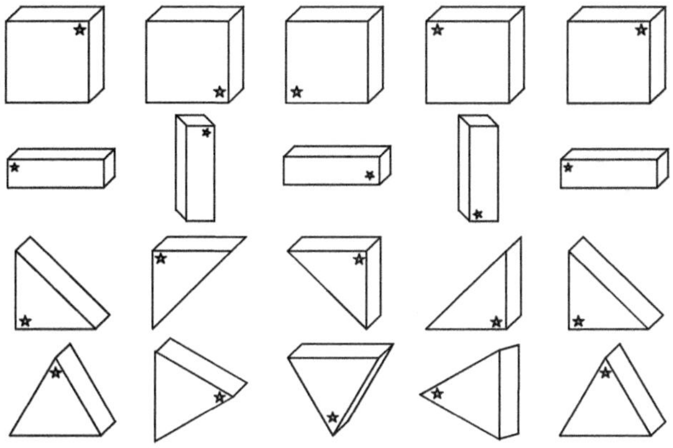

The Language of Maths
Results of the investigation will vary.

Page 2

Let's Investigate
Patterns will vary.

Around the World
The pattern has 180° rotational symmetry.

Around the World
Results of the investigation will vary.

Page 3

Looking for Patterns
Patterns will vary.

Looking for Patterns

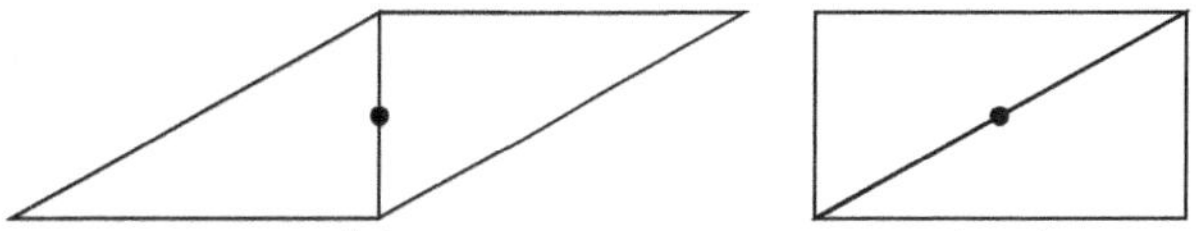

Focus on Science
Results of the investigation will vary.

Page 4

Let's Investigate
Results of the investigation will vary.

The Language of Maths
H, I, N, O, o, S, s, X, x, Z, z

The Language of Maths
Almost all the cards in a pack of playing cards show rotational symmetry. The exact number varies with the design. They are designed to have rotational symmetry so that they can be read from opposite sides of a card table.

Inquisitive ant

anticlockwise
The direction opposite to the one in which the hands of an analogue clock move.

Issue 33

Geometry

Prerequisites for learning

- Describe and identify the position of a square on a grid of squares
- Use coordinates in all four quadrants to draw, locate and complete shapes that meet given properties
- Compare and classify geometric shapes, including quadrilaterals and triangles, based on their properties and sizes; use knowledge of properties to draw 2-D shapes
- Use the eight compass points to describe direction
- Recognise and use quarter turns, both clockwise and anticlockwise and know that a right angle represents a quarter turn
- Know angles are measured in degrees (°)
- Identify:
  - angles at a point and one whole turn (total 360°)
  - angles at a point on a straight line and $\frac{1}{2}$ a turn (total 180°)
  - other multiples of 90°
- Calculate and compare the area and perimeter of rectilinear shapes

Resources

pencil and paper
Resource sheet 2: My notes (optional)
Resource sheet 3: Pupil self assessment booklet (optional)
Resource sheet 11: Coordinates grids (0 to ±4)
Resource sheet 18: 1 cm squared paper
ruler
street directory and / or map of the local area
calculator (optional)

Teaching support

Page 1

Around the World

- The important aspect of this activity is the justifications children give for the position they suggest to locate the post box.

Around the World

- You may wish to tell the children that generally post-people deliver mail to one side of the road first and then the other.

At Home

- When children have written their set of instructions and drawn their accompanying map, they give either their instructions or their map to another child. Can they draw a map from the set of instructions or write a set of instructions from the map?

Page 2

Construct

- Tell the children that there are eight different possible rectangles.
- How many different sized rectangles are there?

Let's Investigate

- If children have difficulties finding the fourth vertex of the kite, suggest that they draw a line between points (1,2) and (–2,1) and then draw another line from point (–1,3) downwards, bisecting the first line. Where this second line crosses a point, this point will make the fourth vertex of a kite.

Construct

- Suggest that children position Laura's home towards the bottom left-hand corner of the grid.

Page 3

Around the World

- Ensure children understand how grid references work on a map or street directory.

Around the World

- Encourage the children to be as accurate in their scale drawing as possible.

Looking for Patterns

- Ensure children realise that, for each angle, they are measuring in a clockwise direction.
- Remind children that one whole turn is 360°, a half turn is 180° and a quarter turn 90°.

Page 4

Around the World

- The directions of north, south, east and west are referred to as the four *cardinal directions* or *cardinal points*. The intermediate points between the four cardinal directions are north-east (NE), north-west (NW), south-west (SW), and south-east (SE) and are referred to as the four *ordinal* (or *intercardinal*) *directions*. There are a further eight divisions on the compass rose. It is these further eight divisions that this activity focuses on.
- Children may get the order of the three letters of each of these additional eight directions mixed up. If so, explain that the first letter is the letter of the adjacent cardinal point (N, S, E, W) and the next two letters are the letters of the adjacent ordinal point (NE, SE, SW, NW).

Looking for Patterns

- Ensure children understand that if you move between any two adjacent points on a sixteen-point compass rose, the size of the angle is 22·5°.

AfL

- Describe to me your recommendations. Why do you think this is the best option?
- Tell me your set of instructions. How does your map complement your instructions?
- Tell me the coordinates for the vertices of this shape. What patterns do you notice in the coordinates?
- If you were to continue this curve what would be the next set of coordinates?
- How do you read a map in a street directory?
- What can you tell me about the 8 / 16 points of the compass?
- What is the size of the angle between these two compass points if you move in a clockwise direction? How did you work that out?
- What is the size of the angle as you move between two adjacent points on an 8-point / 16-point compass rose?

Answers

Page 1

Around the World
Answers will vary.

Around the World
Routes will vary.

At Home
Routes and maps will vary.

Page 2

Construct
(–2,2), (3,2) Perimeter = 12, Area = 5
(–2,3), (3,3) Perimeter = 14, Area = 10
(–2,4), (3,4) Perimeter = 16, Area = 15
(–2,0), (3,0) Perimeter = 12, Area = 5
(–2,–1), (3,–1) Perimeter = 14, Area = 10
(–2,–2), (3,–2) Perimeter = 16, Area = 15
(–2,–3), (3,–3) Perimeter = 18, Area = 20
(–2,–4), (3,–4) Perimeter = 20, Area = 25

Let's Investigate
Point (1,–3) is the fourth vertex of the kite.

Point (0,0) will make the fourth vertex of a square.
Points (2,–1) or (4,–2) will make the fourth vertex of a trapezium.
Almost any point on the grid will make the fourth vertex of an irregular quadrilateral.

You cannot make a quadrilateral that does not have two adjacent sides equal and one right angle.

Construct
Maps will vary.
Descriptions of the walk home will vary but should be similar to the following:
- Come out of the school and turn left.
- Turn right at the intersection, and right again at the traffic lights.
- At the library turn left, walk along the road until you get to a big tree where you turn left again.
- When you reach the post box, turn right. Home is on the left.

Page 3

Around the World
Results of the investigation will vary.

Around the World
Maps will vary.

Looking for Patterns

E and S – 90°	SE and N – 225°
N and S – 180°	NE and S – 135°
N and NE – 45°	N and NW – 315°
SW and NW – 90°	W and E – 180°
E and NW – 225°	SW and W – 45°
W and SW – 315°	N and SE – 135°

The differences between points are in increments of 45°.

Page 4

Around the World

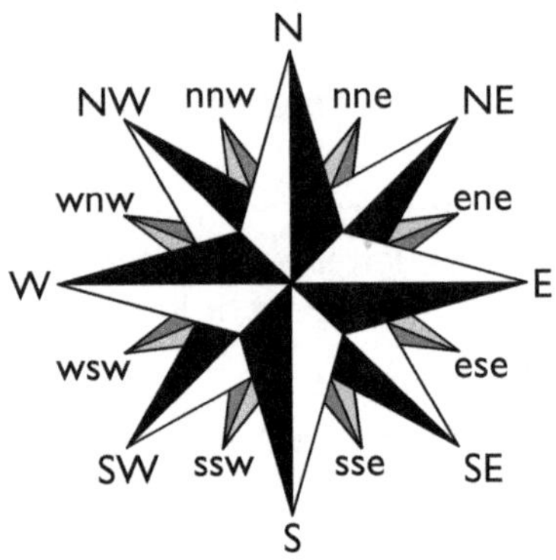

Looking for Patterns

nne and wnw – 270°	wnw and nne – 90°
ssw and sse – 315°	W and ene – 157·5°
ese and sse – 45°	SE and ssw – 67·5°
N and wsw – 247.5°	S and NE – 225°
E and ssw – 112·5°	ese and NE – 292·5°
ene and wsw – 180°	ssw and nnw – 135°
nnw and NW – 337·5°	N and ssw – 202·5°

The differences between points are in increments of 22·5°.

Inquisitive ant

bearing
The location or direction of movement calculated using a map or compass.

Issue 34

Geometry

Prerequisites for learning

- Compare and classify geometric shapes, including quadrilaterals and triangles, based on their properties and sizes; use knowledge of properties to draw 2-D shapes
- Identify, describe and represent the position of a shape following a reflection, translation or rotation, using the appropriate language, and know that the shape has not changed

Resources

pencil and paper
Resource sheet 2: My notes (optional)
Resource sheet 3: Pupil self assessment booklet (optional)
Resource sheet 13: Easter egg design
Resource sheet 14: Coordinates grids (0 to ±12)
Resource sheet 15: Triangular grids
Resource sheet 16: Square grids
Resource sheet 17: Circular grids
Resource sheet 21: Squared dot paper
ruler
examples of cross-stitching (optional)
A4 paper
coloured pencils
set of 1–8 digit cards (optional)
geoboard and elastic bands (optional)
6 paper squares (per child), ideally each a different colour
interlocking cubes
computer with Internet access

Teaching support

Page 1

Around the World

- Children investigate Ukrainian Easter Eggs (Pysanky) commenting on their geometric designs.

The Puzzler

- Children try to solve the puzzle using six, seven, eight and nine disks.
- Children work out a set of rules for finding the solution to the puzzle for any number of disks, i.e.
  For an even number of disks:
  – make a move from peg A to peg B or vice versa
  – make a move from peg A to peg C or vice versa
  – make a move from peg B to peg C or vice versa
  – repeat until complete.
  For an odd number of disks:
  – make a move from peg A to peg C or vice versa
  – make a move from peg A to peg B or vice versa
  – make a move from peg B to peg C or vice versa
  – repeat until complete.
- Discuss with the children the formula for finding the total number of moves required to solve a Tower of Hanoi puzzle with any number of disks, i.e. $2^n - 1$, where n is the number of disks.

Page 2

Looking for Patterns

- Children can extend the patterns in their designs by using colour. Can they use colour to show examples of reflective and rotational symmetry?

The Puzzler

- Introduce children to the Eight Puzzle Problem. Using the digit cards 1 to 8 arranged in a random order on a 3 × 3 grid, the object of the puzzle is to place the tiles in either of the following orders by making sliding moves that use the empty space.

	1	2
3	4	5
6	7	8

1	2	3
8		4
7	6	5

Page 3

Looking for Patterns

- Children can create their quincunx design using ICT.

Let's Investigate

- If possible, show children an example of cross-stitching.
- If appropriate, children can create their cross-stitch needlework pictures or designs using ICT.

Let's Investigate

- Provide the children with geoboards and elastic bands.

Page 4

Construct

- If appropriate, allow the children to work in pairs or groups to make their own origami star.
- Children can investigate other origami shapes.

The Language of Maths

- A rectangle is a quadrilateral with four right angles.
- There are two types of rectangles: squares and oblongs.
- Squares are rectangles in which all sides are of equal length.
- Oblongs are rectangles in which sides are of unequal length.
- Using the terms *square rectangles* and *oblong rectangles* may be helpful for some children in understanding the differences between a square, a rectangle and an oblong.

AfL

- Describe to me how you made your design / pattern. Does it show examples of transformations? How?
- How did you work out the solution to this problem / puzzle? Did you find more than one solution?
- Did starting with an easier pattern first help you to find a solution to the more difficult one?
- What is the rule of the pattern?
- How are these two shapes congruent?

Answers

Page 1

Around the World
Designs will vary.

The Puzzler
Answers will vary.

One possible solution for the three disk Tower of Hanoi is:

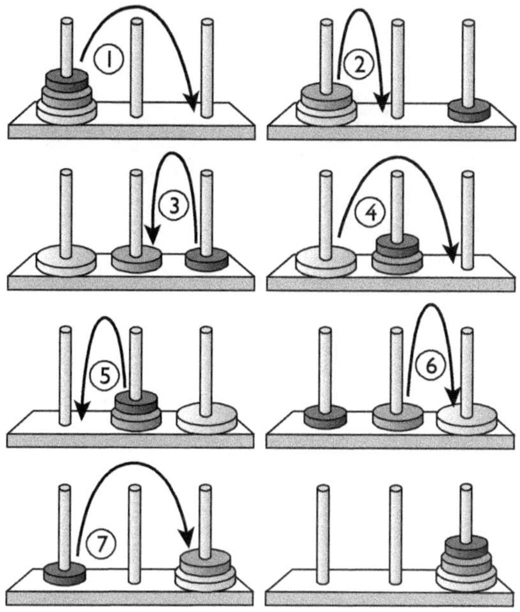

Page 2

Looking for Patterns
Answers will vary.

The Puzzler
The cube can be moved to the empty space in 13 moves.

Page 3

Looking for Patterns
5 as represented on a 1–6 dot dice, playing card and domino all form a quincunx.

Patterns will vary.

Let's Investigate
Results of the investigations will vary.

Let's Investigate
Results of the investigations will vary.

Page 4

Construct
No answer required.

The Language of Maths
A square is a special rectangle with four equal sides.
An oblong is a rectangle which is not a square.

Inquisitive ant

node
A point where two or more lines intersect.

Issue 35

Statistics

Prerequisites for learning

- Read, choose, use and record standard metric units to estimate and measure length
- Interpret and present discrete and continuous data using appropriate graphical methods, including bar charts
- Solve comparison, sum and difference problems using information presented in bar charts, tables and other graphs
- Write a short report of a statistical enquiry, justifying choices and conclusions

Resources

pencil and paper
Resource sheet 2: My notes (optional)
Resource sheet 3: Pupil self assessment booklet (optional)
Resource sheet 18: 1 cm squared paper
Resource sheet 20: 0·5 cm squared paper
ruler
simple PE equipment
TV schedules
computer with Internet access
data handling software

Teaching support

Page 1

Around the World

- As this activity is quite open-ended, you may wish to ask the children to investigate the cost and duration of crossing for travelling only during the summer months.
- Alternatively, you could ask the children to investigate the cost and duration of crossing at any time throughout the year between the same two ports, e.g. Dover and Calais.
- Suggest children present their results in a table.

Let's Investigate

- There is a correlation between foot size and height.
- If appropriate, children create their table using ICT.
- There is a closer correlation between foot size and limb size than there is between foot size and overall height. Unless something has interrupted normal patterns of growth, then the length of the foot from heel to tip of the big toe is exactly the length of the forearm between the inside of the elbow and the bend of the wrist. Ask the children to investigate this.

The Language of Maths

- This activity highlights the varying interpretations that different people can have of the same set of data. What is important, however, is that children are able to give a reasonable explanation as to what the bar chart might represent, and recognise and logically explain the trends that the chart displays. Therefore once the children have completed the activity, conduct a group discussion providing an opportunity for each child to describe their interpretation of the bar chart.
- If appropriate, children write their explanation using ICT.

Page 2

Famous Mathematicians

- Ensure children understand how doublets work. If necessary, work through other examples with the children.

BUS	MICE	HEAD
BUT	MITE	HEAL
BAT	MATE	TEAL
HAT	MATS	TELL
	RATS	TALL
		TAIL

Looking for Patterns

- The paper is more likely to fall face up than down; however, the greater the height it is dropped from, the less likely this is to be true. The purpose of the activity is for children to think about how to record and organise the data.

Page 3

Sports Update

- This activity is best spread out over a week or longer. This will give the children greater opportunity to practise the activity and therefore hopefully to improve their skill level.

Sports Update

- Given that most sports, including football, are played at the weekend, rather than investigating the schedules of sports programmes for just one day, children look into the amount of sport that is televised during an entire week.
- Children can use the Internet to find out TV schedules. If appropriate, they can also write their report using ICT.

Page 4

The Arts Roundup

- Children may re-represent the data in the bar chart as a table, pictogram, block graph or bar line graph.
- Children either draw the table / graph or use ICT.

The Arts Roundup

- The important aspect of this activity is the justifications that children offer for the inferences that they make. Ensure the children provide examples from the bar chart to substantiate their statements.
- If appropriate, children write their report using ICT.

AfL

- Explain to me the results of your investigation.
- What conclusions can you draw?
- What can you tell me about the information displayed in the graph / chart / table?
- What do you think this chart represents? Why do you think this?
- What relationships / connections / trends do you notice?
- Tell me about your experiment.
- Why did you decide to present your data in this way? How else could you have displayed the data?

Answers

Page 1

Around the World
Different routes include:
Harwich to Hook.
Dover to Dunkirk and Calais.
Folkestone to Calais.
Portsmouth to Le Havre, Caen and Cherbourg.
Poole to Cherbourg and St. Malo.
Plymouth to Roscoff.
Costs vary depending on ferry company and time of year.
Time taken varies depending on type of ferry and route.

Let's Investigate
Results will vary.

The Language of Maths
Answers will vary.

Page 2

Famous Mathematicians

cold	fire	hand
cord	hire	band
word	here	bond
worm	herd	fond
warm	head	food
	heat	foot

Other answers are possible.

Looking for Patterns
Results will vary.

Page 3

Sports Update
Results will vary.

Sports Update
Results will vary.

Page 4

The Arts Roundup
This could also be shown as a pie chart, showing each day's contribution to the weekly ticket sales.

The Arts Roundup
Answers will vary.

Inquisitive ant

$\in$
To belong to or to be a member of a particular group or set; i.e. to be an element of.

Issue 36

Statistics

Prerequisites for learning

- Interpret and present discrete and continuous data using appropriate graphical methods, including bar charts and line graphs
- Solve comparison, sum and difference problems using information presented in bar charts, tables and other graphs
- Write a short report of a statistical enquiry, justifying choices and conclusions

Resources

pencil and paper
Resource sheet 2: My notes (optional)
Resource sheet 3: Pupil self assessment booklet (optional)
Resource sheet 18: 1 cm squared paper
Resource sheet 20: 0·5 cm squared paper
ruler
stopwatch
collection of six different cereal boxes
collection of different daily and weekend newspapers
computer with Internet access
data handling software

Teaching support

Page 1

Sports Update

- Children between one and ten years old can have resting pulse rates between 70 and 120 beats per minute (bpm).
- Children older than ten normally have a pulse rate between 60 to 100 bpm.
- Girls tend to have slower pulse rates than boys.
- It is expected for pulse rates to increase by 10 to 30 bpm during exercise.
- Children either draw the graph or use ICT.

Let's Investigate

- You may wish to give the children a list of the top ten television ratings. These can be found either on the Internet or in many of the Sunday newspapers.
- Children can undertake this activity in different ways. One way may be to ask individual children to survey the viewing habits of their family only. Children can then pool their results and draw conclusions about the viewing habits of the families for the entire class.
- Alternatively, children could work together to design a survey to give to a sample of the other children in the class for them to gather information about their family's television viewing habits. Children can then collate these results and draw conclusions about the viewing habits of the families for the entire class.
- If appropriate, children represent the data using ICT.

Page 2

At Home

- Once the children have completed the activity, ensure that there is an opportunity in class for pairs or groups of children to discuss their results.

Let's Investigate

- Children need to be made aware that their proportions will be based on approximations, and that their conclusions will therefore be generalisations.

Let's Investigate

- There are two different aspects to this investigation. Firstly, drawing the graphs and secondly using this information to write a report about the healthiest cereal.
- Children either draw the graphs or use ICT.

Page 3

Sports Update

- For someone who visits the fitness centre twice a week, the difference between paying £5 per visit or £40 per month is minimal. For those visiting more frequently it is more expensive to pay per visit. Therefore, if children have different answers, you may wish to discuss with them their reasoning.

Let's Investigate

- The important aspect of this investigation is the designing of the data collection sheet. Children need to think very carefully about what they need to include on their sheet.
- Children design their data collection sheet and write their report using ICT.
- Once the children have completed the activity, it may be worthwhile holding a group discussion to evaluate the strengths and weaknesses of the different data collection sheets.

The Language of Maths

- Ensure children follow each of the five steps in this data handling investigation.
- If appropriate, children represent the data using ICT.
- Once the children have completed the activity, hold a group discussion providing an opportunity for children to talk about what they did in each of the five steps.

Page 4

Around the World

- Ensure children are able to read and interpret bar charts and line graphs.
- The important aspect of this activity is the rationales that children give for their recommendations based on the data provided.

AfL

- Explain to me why you made those recommendations. What are your recommendations based on?
- Tell me the results of your investigation / survey.
- Talk me through how you drew your chart / graph.
- Why did you present the information in this way? Why did you think it was the best way?
- What do these graphs / charts show? What conclusions can you draw from them? What predictions can you make? How can you justify your predictions?
- Why is this data best presented in this way? What benefits does it have over other types of graphs / charts?

Answers

Page 1

Sports Update
Graphs will vary.

Let's Investigate
Results of the investigation will vary.

Page 2

At Home
Results of the investigation will vary.

Let's Investigate
Results of the investigation will vary.

Let's Investigate
Graphs and reports will vary.

Page 3

Sports Update
7 days a week: Monthly subscription
5 days a week: Monthly subscription
Twice a week: Monthly subscription
Once a week: Pay per visit

Let's Investigate
Results of the investigation will vary.

The Language of Maths
Results of the investigation will vary.

Page 4

Around the World
Reports and rationales will vary.

Inquisitive ant

histogram
Similar to a bar chart. However, in a bar chart all the bars are the same width and it's the height of the bars that is important. In a histogram, both the width and height of the bars are important, i.e. the total area of the bar represents the value.

Stretch and Challenge 5 Record of completion

Class/Teacher: ____

Domain(s)	Topic	*Stretch and Challenge* Issue	Names									
Number: – Number and place value	Number	1										
	Number	2										
	Number	3										
	Negative numbers	4										
Number: – Addition and subtraction	Addition	5										
	Subtraction	6										
Number: – Multiplication and division	Multiplication	7										
	Division	8										
Number: – Addition and subtraction – Multiplication and division	Mixed operations	9										
	Mixed operations	10										
	Mixed operations	11										
	Mixed operations	12										
Number: – Fractions (including decimals and percentages)	Fractions	13										
	Fractions	14										
	Decimals	15										
	Percentages	16										
	Percentages	17										
	Fractions, decimals and percentages	18										

Domain(s)	Topic	*Stretch and Challenge* Issue	Names									
Measurement	Length	19										
	Mass	20										
	Volume and capacity	21										
	Time	22										
	Temperature	23										
	Area	24										
	Area and perimeter	25										
	Measurement	26										
Geometry: – Properties of shapes	2-D shapes	27										
	3-D shapes	28										
	Movement and angle	29										
Geometry: – Position and direction	Reflections	30										
	Translations	31										
Geometry: – Properties of shapes – Position and direction	Reflective and rotational symmetry	32										
	Geometry	33										
	Geometry	34										
Statistics	Statistics	35										
	Statistics	36										

S&C Volume 5

The Maths Herald

Name: Date:

Date of starting issue:__________ Date of finishing issue: __________

My notes

My notes

Inquisitive ant

2

My notes

3

My notes

S&C Volume 5

The Maths Herald

Name: Date:

Date of starting issue: ____________ Date of finishing issue: ____________

What have you learned?

What did you use to help you?

Any other comments?

Your teacher's comments

1

4

What did you enjoy the most?

What did you enjoy the least?

Did you find the work:

too easy? ☐

just about right? ☐

too hard? ☐

What would you like to learn next about this topic and why?

9 x 9 grids

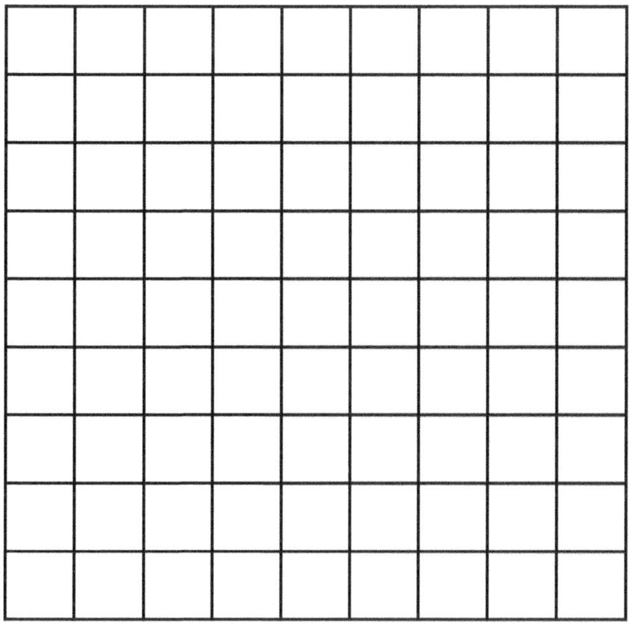

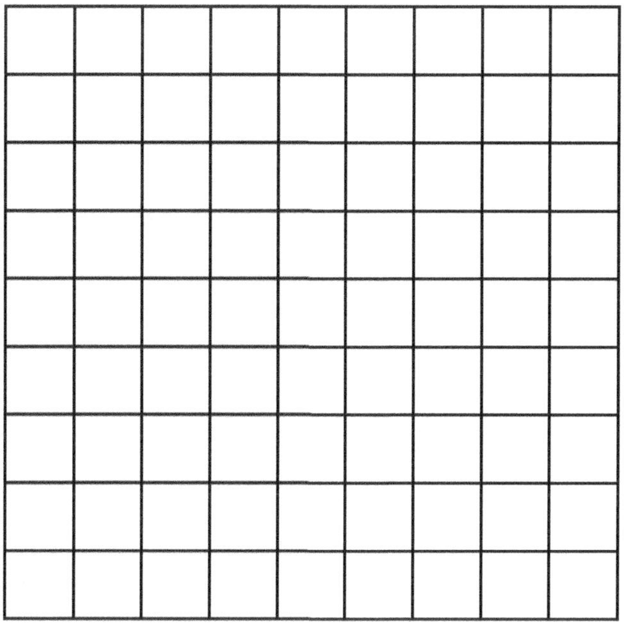

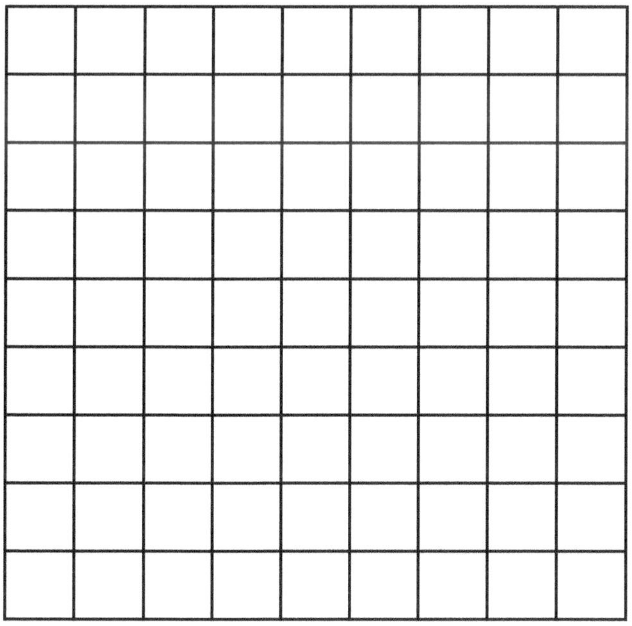

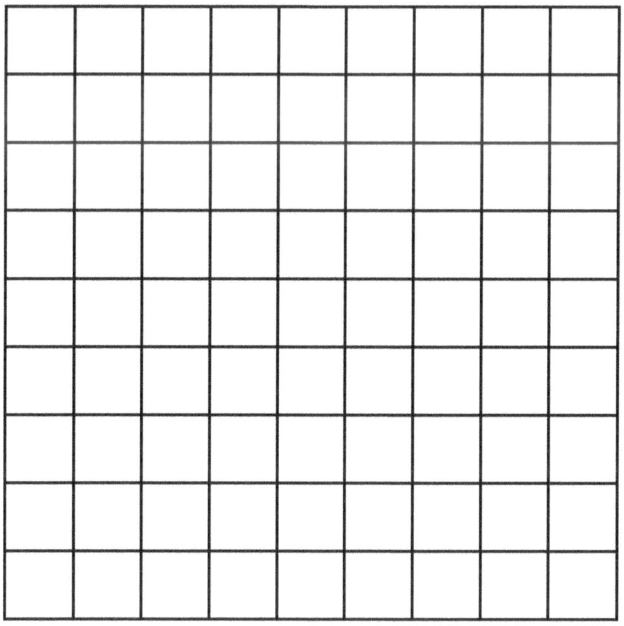

Music paper

Pie charts

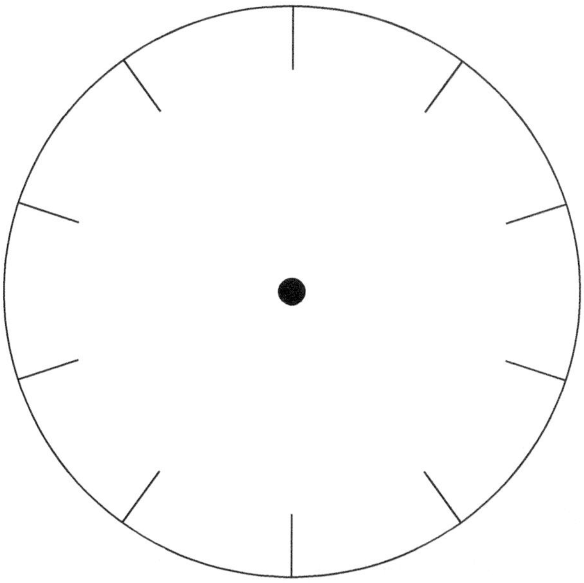
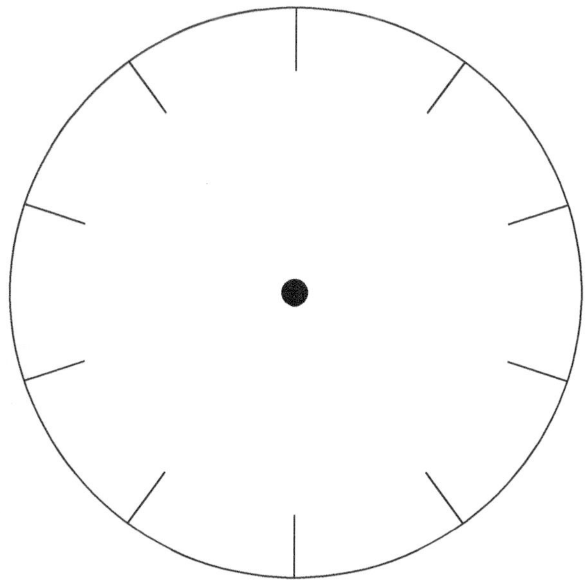
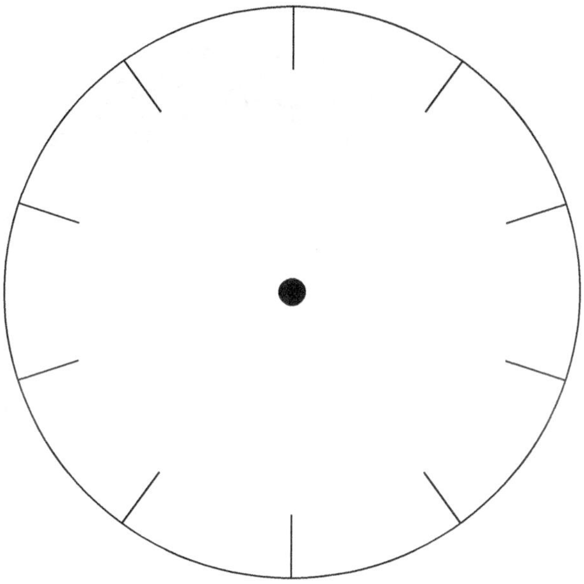
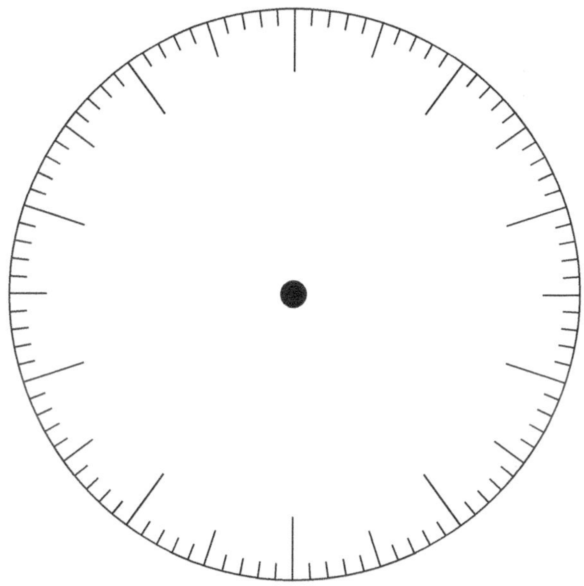
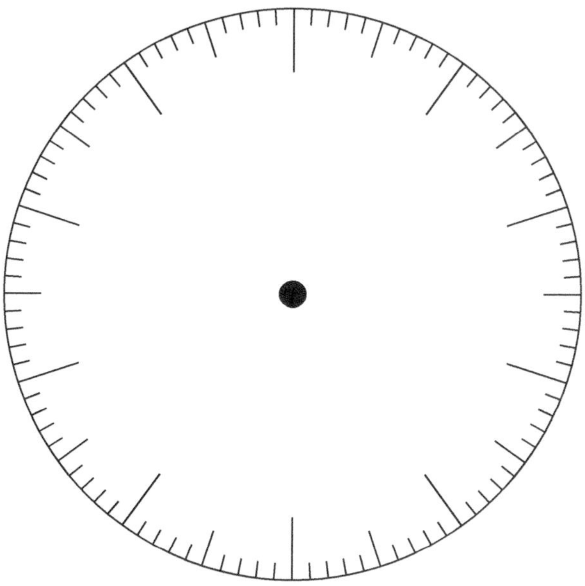
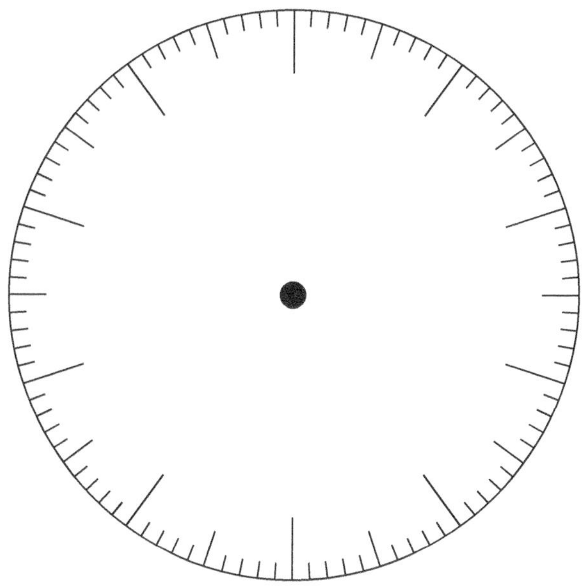

Pythagoras's Theorem

Seven circles patterns

Analogue clocks

Polygons

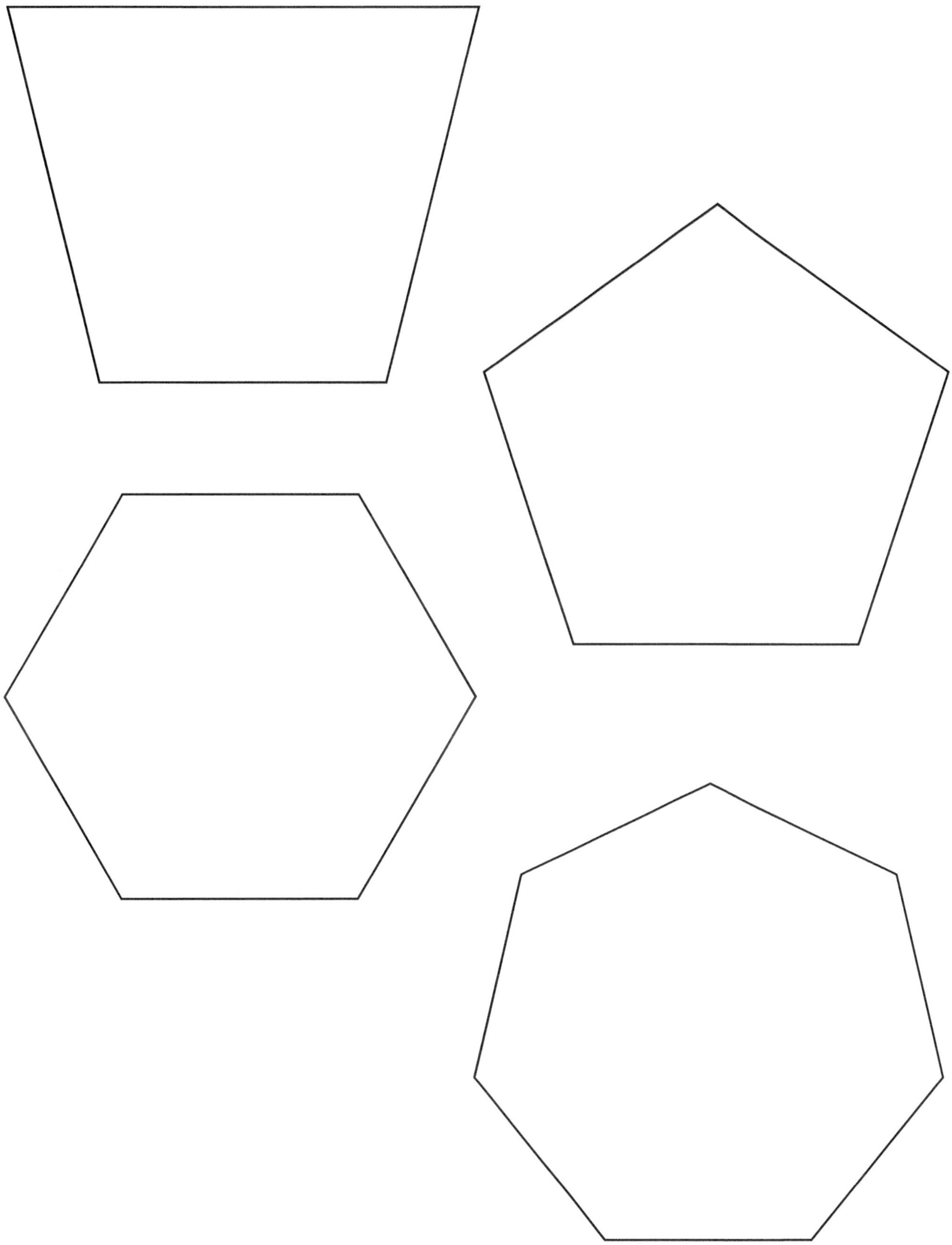

Coordinates grids (0 to ±4)

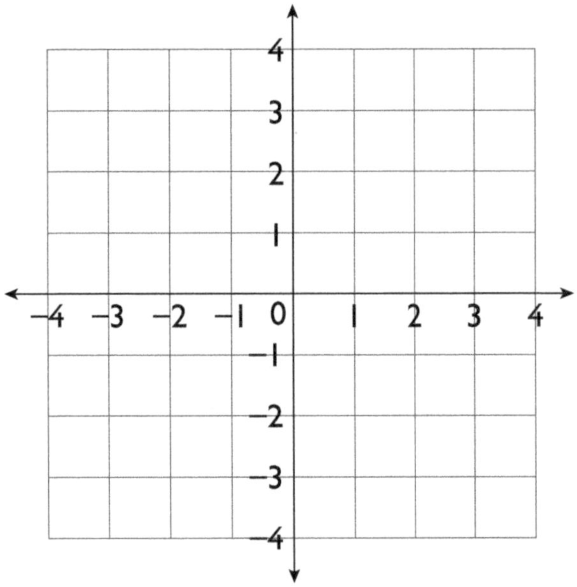

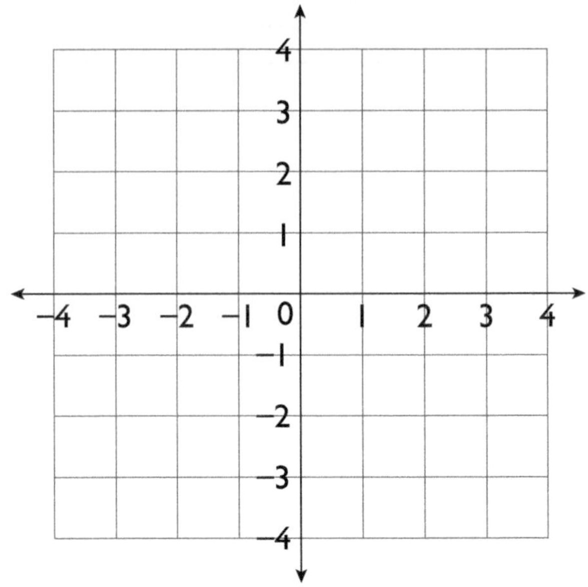

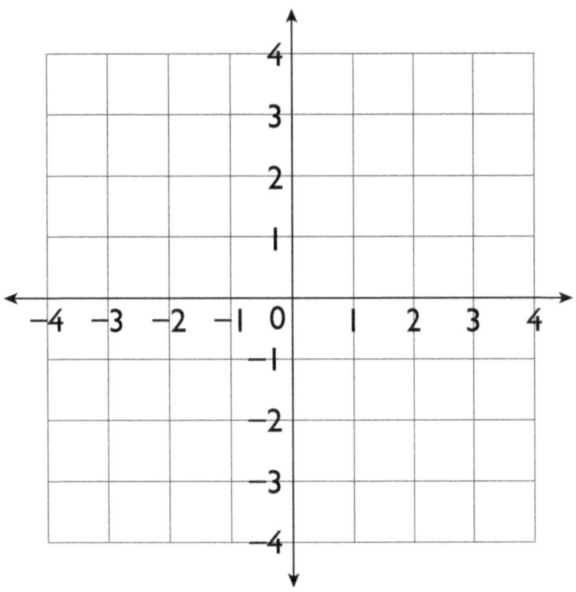

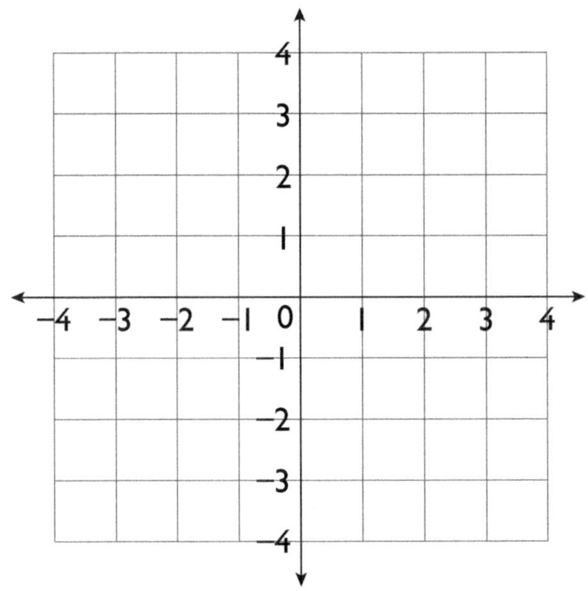

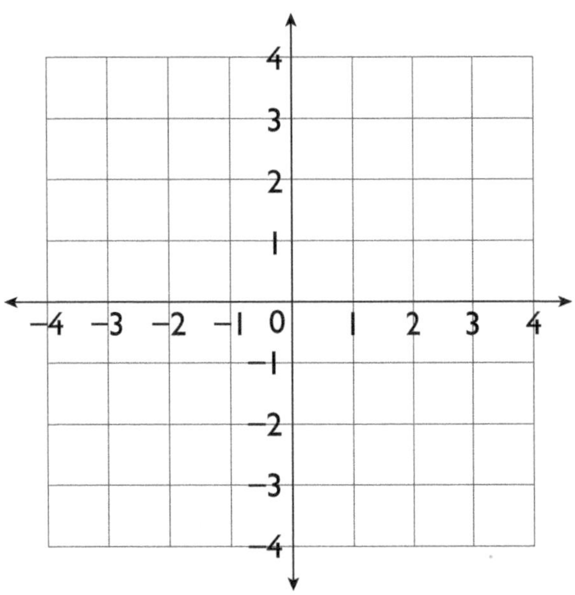

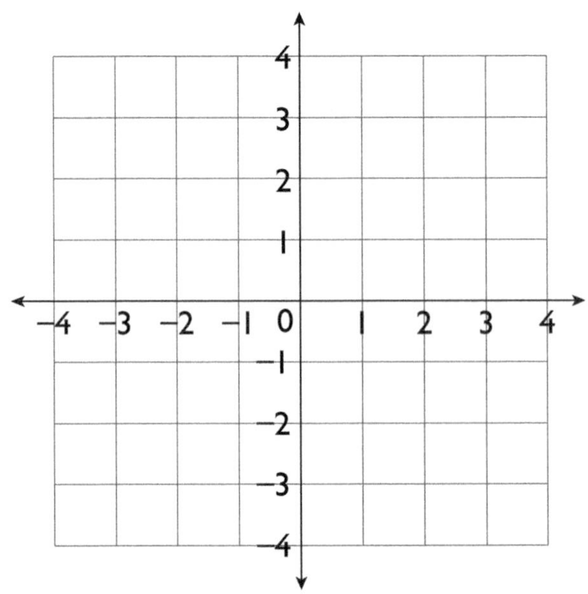

Translating squares

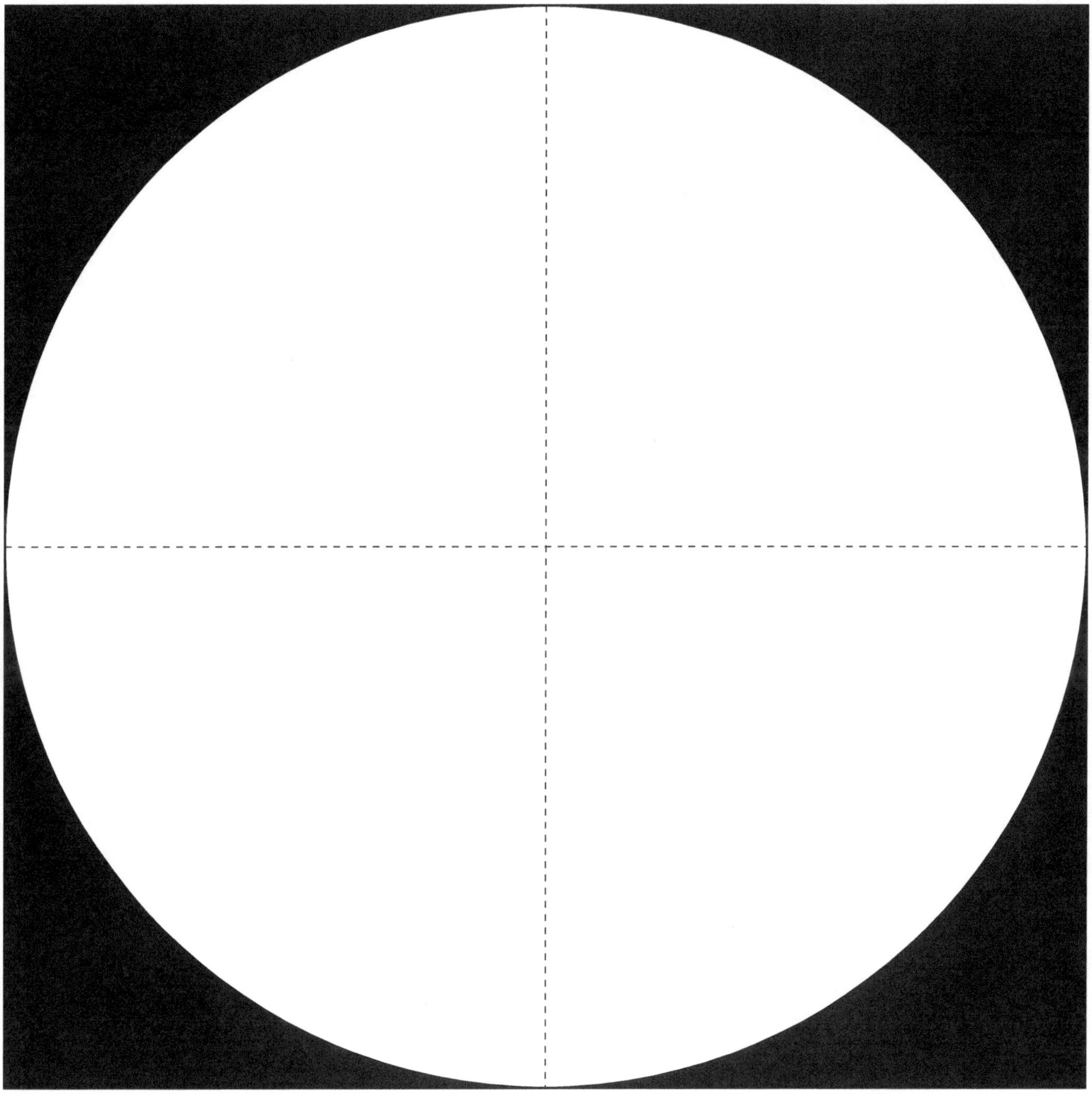

Easter egg design

Coordinates grids (0 to ±12)

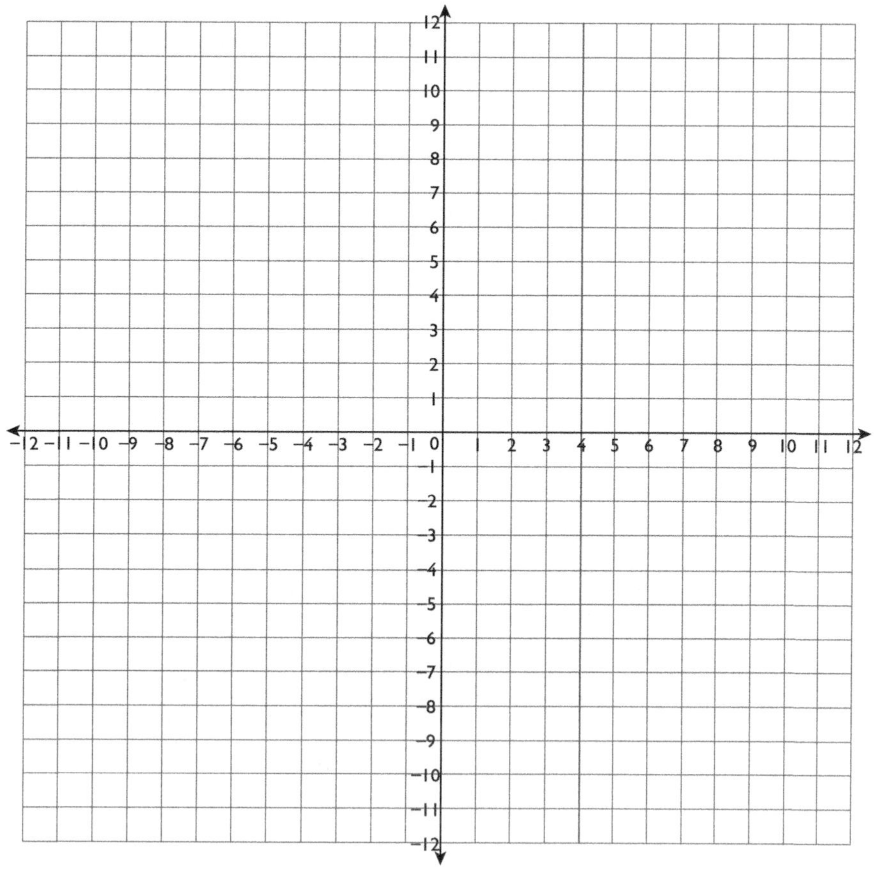

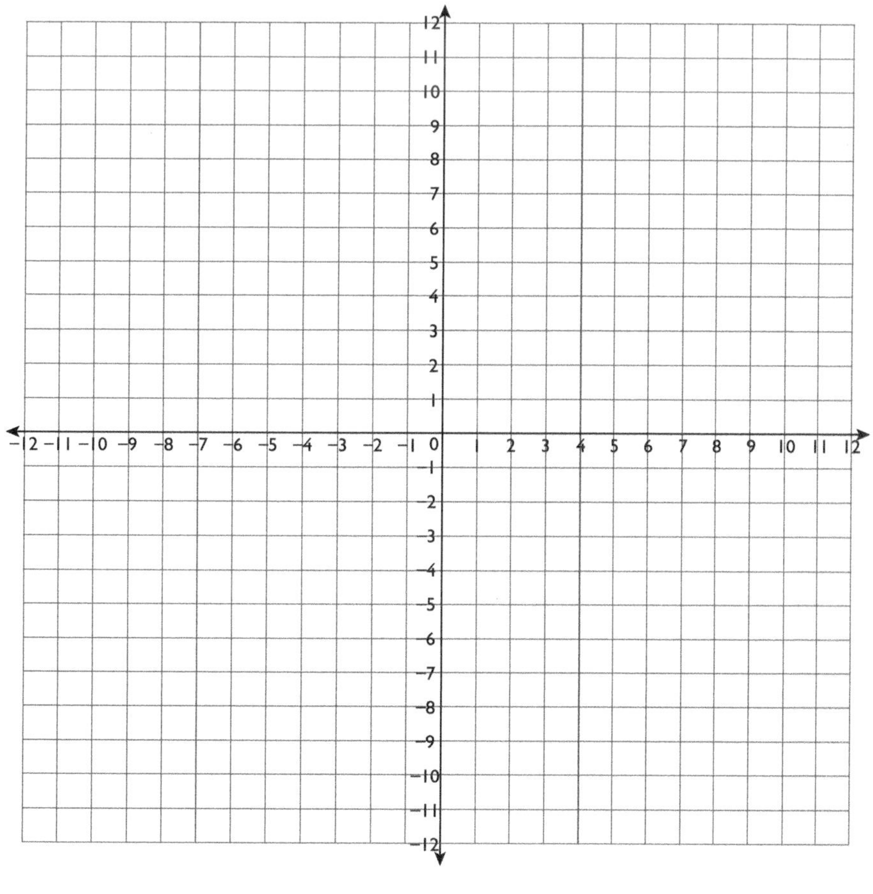

Triangular grids

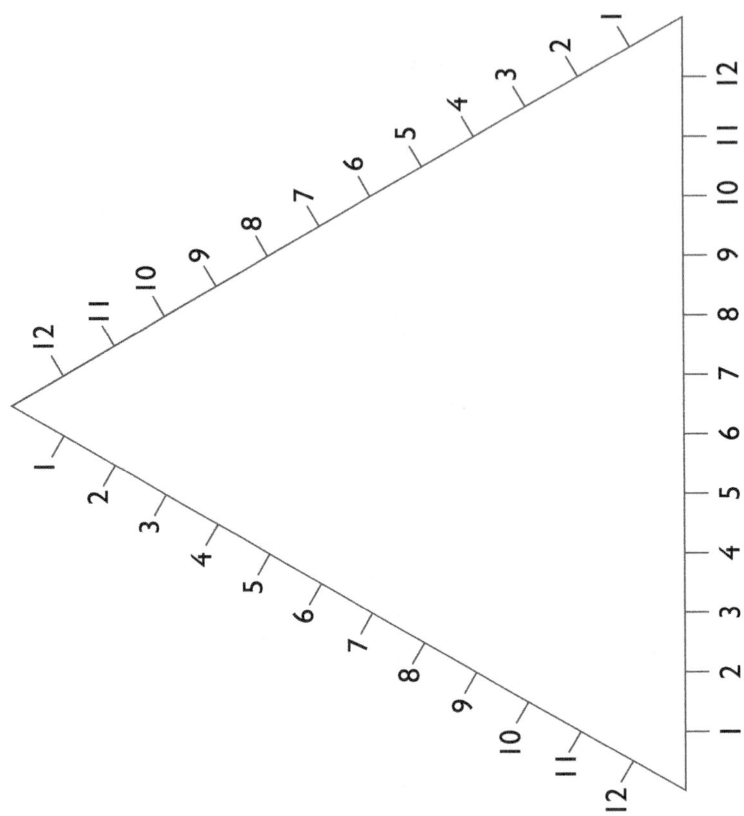

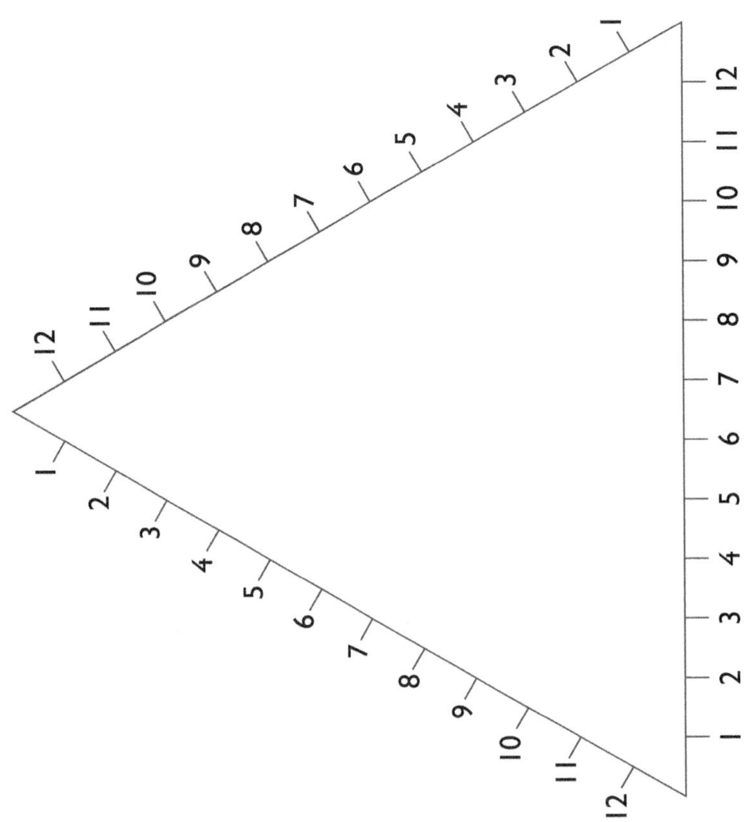

Square grids

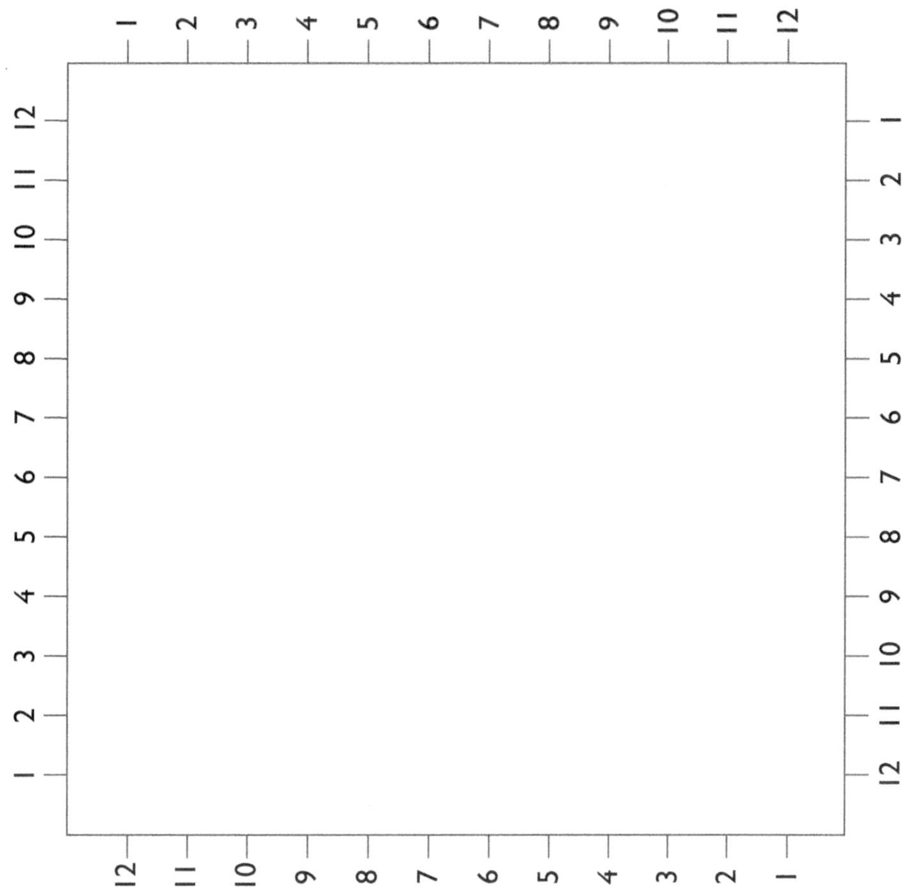

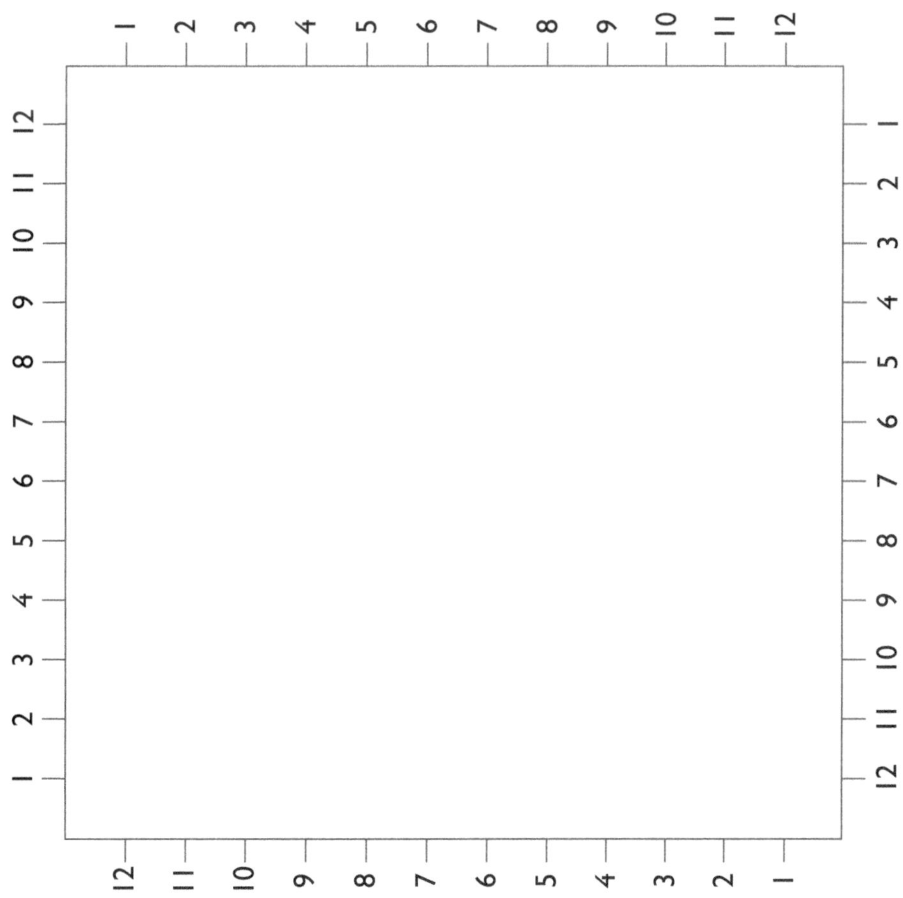

Circular grids

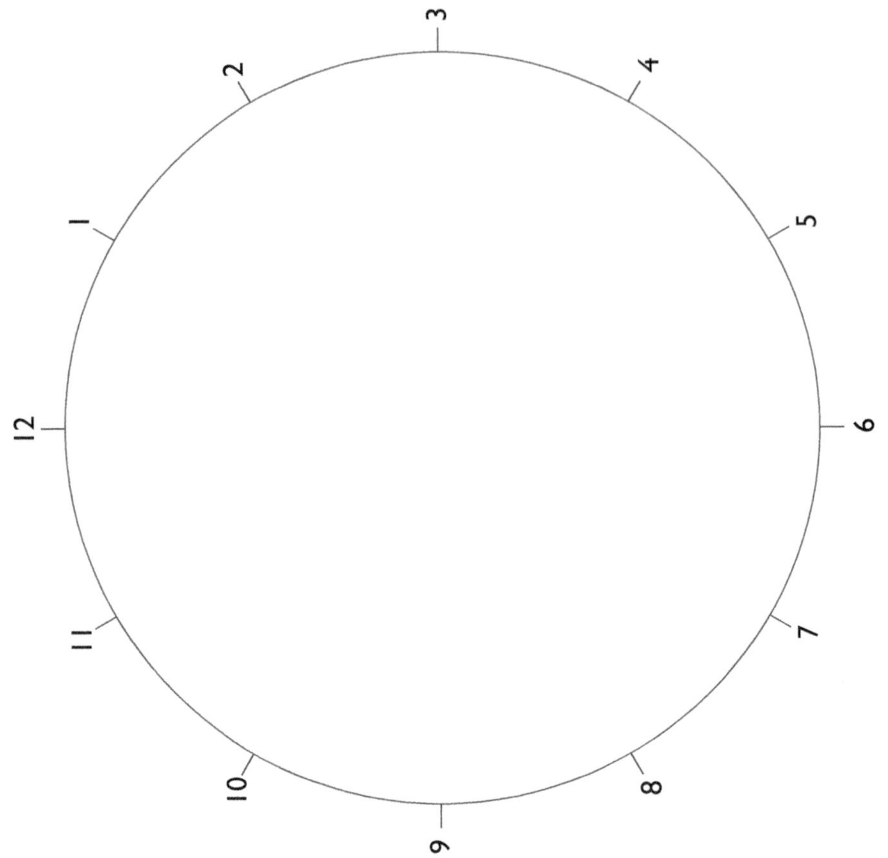

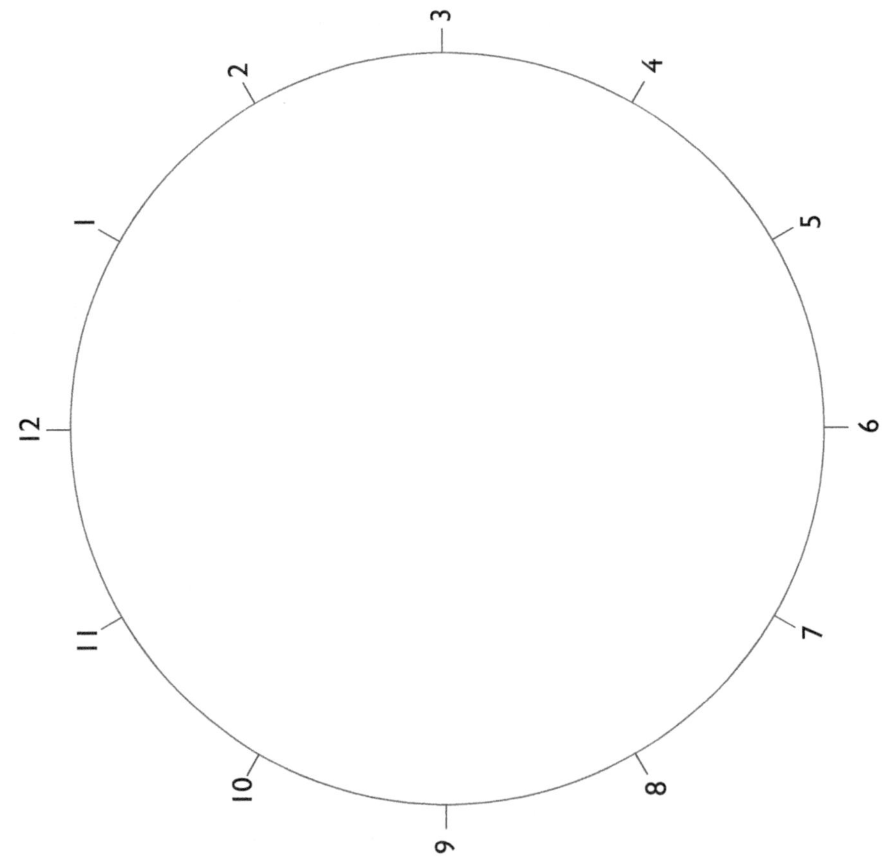

1 cm squared paper

2 cm squared paper

0·5 cm squared paper

Squared dot paper

Triangular dot paper

Isometric paper

Diamond paper

Notes

Notes

Notes

Notes

Notes